The Internet We Could Have Had

The Internet We Could Have Had

CHRISTOPHER M. KELTY

polity

The publication of this book was supported by a grant from the Livescu Foundation and the Livescu Initiative for Neuro, Narrative and AI at the University of Los Angeles, California.

NNAI
LIVESCU

First published in 2026 by Polity Press

Polity Press
65 Bridge Street
Cambridge CB2 1UR, UK

Polity Press
111 River Street
Hoboken, NJ 07030, USA

ISBN-13: 978-1-5095-7450-6

A catalogue record for this book is available from the British Library.

Library of Congress Control Number: 2026931182

Typeset in 11 on 14 pt Warnock Pro
by Cheshire Typesetting Ltd, Cuddington, Cheshire
Printed and bound by CPI Group (UK) Ltd, Croydon

For further information on Polity, visit our website:
politybooks.com

For Hannah, unconditionally

Contents

I 404 Not Found

> Since the end of human action, as distinct from the end products of fabrication, can never be reliably predicted, the means used to achieve political goals are more often than not of greater relevance to the future world than the intended goals.
> – Hannah Arendt, *The Human Condition*[1]

Once upon a time I was writing a different book about the internet. It was full of hackers, piracy, openness, decentralization, free speech, anarchist activism, and the war over intellectual property rights. It was about the configuration of the technical networks that would form the future of political life, and the possibility of liberation. It was somewhere in the years of the Obama presidency, in that moment between the very different deaths of Steve Jobs and Michael Brown.

The Great Recession had stoked the Occupy movement, the Arab spring and the Indignados/M15 movements; Chelsea Manning and Wikileaks had shaken up the US military establishment with revelations of unaccountable violence; pirate parties were gaining seats in European parliaments; decentralized free software technologies were enabling hackers and activists across the Global South and North; #hashtag activist was just emerging and Black Twitter was just being theorized; Aaron Swartz's suicide and the SOPA/PIPA legislation galvanized protests and anger against the giant media conglomerates who anxiously swatted at both everyday downloaders and flamboyant charlatans like Kim Dotcom. Among it all, Anonymous was daily and heroically engaged in, as they used to say, various internet motherfuckery.

If none of this sounds familiar, then it is likely you were paying attention to something else entirely at the time; perhaps the internet was just another utility bill, or a mysterious place of promises that no one could quite explain, unlike the iPhone which was sleek and obvious. Maybe you were wondering about China's new Xi or India's new Modi; maybe you were trying to convince friends to use Facebook so you would have someone to use Facebook with (or Google+, if you made bad choices), playing Portal 2, grooving on Gangnam Style, or perhaps, out of the corner of your ear, catching the name Edward Snowden on the news.

When Edward Snowden revealed in 2013 that the US National Security Agency, in cooperation with their allies, had been systematically spying on nearly everyone in the world – including their own citizens – via the internet, something changed in public consciousness. Snowden's revelations simultaneously changed the internet from a utopian dream that no one really took seriously, to a dystopian horror that everyone did.[2]

If you are too young to have lived consciously through any of this – like most of my students these days – you live with a different internet: a tool of political domination. All these names are unfamiliar or irrelevant in this day and age of AI and technocratic recklessness everywhere. Attention-sucking social media, algorithmic supplication, giant platforms, racial bias, the internet of things, big data, the blockchain, cookies, the end of privacy, anti-trust law battles, hate speech, misinformation, white nationalism, authoritarianism, and Donald

Trump memes. It is capitalism, colonialism, imperialism, slavery, and environmental destruction all rolled into one hideous hydra whose heads are Zuckerberg, Bezos, Pichai, Cook, Nadella, with Musk and Thiel at the ass end shitting out Sam Altman.

The internet of now couldn't be more different than the one I had been writing about. I would never finish writing that book because somewhere in these years I had become, as my wife pointed out to me, a scholar of *the internet we could have had.*

The internet we could have had is a phrase that resonates for many people. And for a wide variety of reasons, too. After all, the internet is undefinable; it is everything and more, all those memes, all those protocols, all those TikTok videos, all the wi-fi routers and browsers, all that shopping, all those web forms and constant attempts to remember passwords, all that porn. Nonetheless, *the internet we could have had* is something real: it could have been a global brain; a universal library; a decentralized market; a global public sphere; a collective intelligence; an anarchic medium for resistance and connection; a tool of art, joy, and curiosity; a medium of creativity and innovation. Maybe you miss it. Maybe you missed it. Maybe, like my children, you wonder why we have the internet we do and what exactly was wrong with the internet we could have had, and who the actual fuck is responsible?

What happened to the internet we could have had? Can we actually remember a thing that never came to pass? As Hannah Arendt suggests in the passage above, when it comes to making new stuff, sometimes the means are of greater relevance to what could have been than the ends; this is the case for the internet we could have had.

It's not hard to find claims for what happened, or indictments of those responsible. Since about 2017, a "techlash" has deflated the bombastic claims of Silicon Valley tech giants, or tried to, but not their profits or power. For example, this 2022

New York Times op-ed by Ben Tarnoff, author of *Internet for the People*, suggests: "The root [of the problem] is simple: The internet is broken because the internet is a business. . . . Even with the best regulatory and antimonopoly measures, corporations would still own the internet. Immensely consequential decisions would be left in the hands of executives and investors. Most people would have no say in matters that centrally affect their lives."[3]

The internet is no longer "for the people"; it is no longer a public good; it is privatized or stolen. Shoshana Zuboff agrees, at great and hefty length in the widely cited *The Age of Surveillance Capitalism*. Zuboff's book makes a powerful case for the emergence of a new form of "instrumentarian" power (a newer more terrifying totalitarianism) wielded by capitalism to extract and monetize our behavior, stealing from us not just a public good, and not just our time and our privacy, but our very humanity. "Information civilization" is not simply, as she says, a *coup d'état* but a *coup des gens*, an overthrow of popular sovereignty itself.[4]

Or here is Jonathan Crary in the 2022 pamphlet *Scorched Earth*:

> The internet complex is the implacable engine of addiction, loneliness, false hopes, cruelty, psychosis, indebtedness, squandered life, the corrosion of memory, and social disintegration. All of its touted benefits are rendered irrelevant or secondary by its injurious and sociocidal impacts. . . . The narrative now – of an egalitarian technology endangered by monopolistic corporations, the rescinding of net neutrality, and invasions of privacy – is plainly false. There never was or will be a "digital commons."[5]

Case, apparently, closed. It should never have come to exist, and Crary, for one, is clear we would be better off if we got rid of the internet entirely, which is an interesting, if insane, idea.

There are many other examples, more and less ambitious and magisterial in their attempts to account for the power of Big Tech. As these rapacious, unregulated firms arose, they brought with them new colonialisms, new capitalisms, new imperialisms, new racisms. The repetition of *new* here mirrors the specs that Silicon Valley demands of any idea: novelty first, utility second.[6]

Behind the claim of novelty, most of the techlash critiques assert that nothing has changed, which makes them not so useful as theories. Because capitalism is insatiable, the internet never could have been anything other than a tool of extraction and surplus value; because power is absolute, the internet never could have been anything other than the birth of a new empire, consolidated by Wall Street and the US government, at the behest of Silicon Valley. Because colonialism is ongoing, the internet never could have been anything other than a racialized form of wealth transfer from the South to the North. Their never was any "internet we could have had": you were a sucker for ever believing in it in the first place.

It's hard to argue with the idea that the internet has gone to shit. The evidence is pretty clear. Indeed, even one of the most enthusiastic defenders of the internet we could have had, Cory Doctorow, agrees with the facts, if not the prescription: he describes the transition appropriately as "enshittification" – platforms stole everything from us to please business interests, and then they stole everything from those business interests, leaving only a salted plain of destruction.[7]

Across all these diagnoses, however, something is gone. The story of the internet is missing a section, like the 18 minutes of the Watergate tape, or maybe just the drunkards' blackout. It's just not clear how we got here. Even worse, to say that the internet we have today was inevitable, or to assert that we were suckers for believing in a different future – this is a politically disabling thing to say. It makes us all into victims of Big Tech, who stole it from us, leaving only that melancholy error page which sums it all up: "404 Not Found."

Happily, most of these critiques and analyses are not silent on what to do next. If you look to them for a prescription, they are equally ambitious. As Doctorow counsels in *The Internet Con*: "Tech is the terrain on which our future fights will be fought. If we can't seize the means of computation, we will lose the fight before it is even joined."[8] The way to fix the internet we do have is to double down on a more radical participatory democracy, a reform of property rights or the financial system, crush the corporate monopolies, create a just and equal cooperativism with a robust decentralized, federated technology, build community-run and community-governed institutions, with forms of mutual aid or even anarchism.

In short – and this should come as a surprise – they argue for the internet we could have had.

This is a weird puzzle. Why, if the internet we could have had aimed at even attenuated versions of any of these lofty goals, did it give us the internet we do have? Why would we simply repeat this mistake, if in fact it was a mistake?

So here I am, writing another book about the internet. This one is a *forensic* one, an inquiry: was the internet we could have had an innocent victim or a nefarious mastermind? Was it killed, or did it disappear? What was the motive, the time of death, the opportunity, or the weapon?

The internet we could have had was a global, collective experiment in popular power: free speech, participation, openness and anonymity, decentralization and emergence, activism and resistance. These are momentous aspects of individual and collective political life, and the internet we could have had was an opportunity to create new versions of these things, to re-organize or re-assemble them. More importantly, it was possible to do this beyond existing forms of authority and rule: through mundane acts of everyday tinkering and playing; from experiments with animated gifs and HTML to software and markup, semantic webs, net.art, social graphs, hacking and leaking, software licenses, protocols and browsers, cloud computing and business processes. Entrepreneurs, experimenters, engineers, writers, gamers, geeks, hackers all took these things very seriously, not just as hobbies or schemes, but as ways to transform the world completely. Then one day, the internet we could have had was gone and something totally different was in its place.

To make matters worse, in narrating this mystery, I will be a classic unreliable narrator: I was there, I saw the victim last, I could be a suspect. Maybe I believed the hype, and so you shouldn't trust anything I say? *Caveat lector*. How will we recover the evidence about what the internet could have been, about what people said it would be? Who did what, when, why?

My story is about the *action* that people engaged in, the attempts people made to begin something new, or fix something old, which also includes what we *said* internet could have been: many of the things we wrote or theorized about what it would be. It is a pragmatic account, and an ambivalent one. It matters what people did, concretely and empirically, but it also matters what story we tell about it, *ex post facto*, and these are not the same things. The difference in this case is that many people, myself included, thought themselves to be beginning one thing, only to find something very lamentable come to pass. If we were to begin again, would we do the same thing? Can we begin again?

Thus, this is also a *proleptic* book – one that flashes backwards and then forwards again in time in order to ask: What happened when? Were the signs there all along? Proleptic because, in order to diagnose what has gone wrong with the internet we do have (and indeed, what we risk in the AI we could get), we have to respect that it was not a repetition or a mistake: it was a *beginning*, maybe many beginnings, and it is still unfolding. We must look back, from our contemporary moment, hazy with confusions about AI, to this prior moment, and reconnect those threads that might help us understand what happened to the internet we could have had and whether we can do anything about the AI we are making right now.

So what did happen?

The answer – spoiler alert – is that the internet we could have had failed for the same reason it could have succeeded: because of the very power of its openness and decentralization, its modifiability, its extensibility, and its experimental participatory form. Just as the internet we could have had encouraged many different and diverse people to imagine and assemble alternatives, it also allowed them to fail to defend against takeover and appropriation. The internet we could have had failed not because powerful people controlled it, but because no one did. At the heart of the idea of popular power, of *popular sovereignty*, is this conflict: the very ability to put limits on authority is unlimited. The people are sovereign, but to exercise that sovereignty we have to understand how our power works and we need to know what kind of rule we seek to limit. The internet we could have had was a test of this power and its failure should give us serious pause. To understand what went wrong is one goal of this book.

We killed the internet we could have had. Not unlike the outcome of *Murder on the Orient Express*, everyone is the murderer. The internet we could have had was not designed (or killed) by any one person or organization (certainly not by "Silicon Valley") but by all those who participated in making the software, protocols, laws, movements, institutions, theories, and companies. If these people – people everywhere around the earth, and of all types, disputing, negotiating, trading, organizing, and betraying – if these people made the internet we could have had with hopes of better worlds, then why did we get a worse one? Why have the ideas, tools, institutions, and laws intended to liberate and democratize created instead the conditions for something even worse than what they were made to transcend? If the answer is that Big Tech was already so powerful, then we ought to have noticed sooner; if the technology was so obviously destined to immiserate us, why did we embrace it instead of resist it? These questions are not easy to answer, and the simple answers – "capitalism," individual human weakness, fundamental human greed, corporate corruption, lack of regulation, or bad design – don't really get to the heart of the matter. The internet we could have had was a referendum on popular sovereignty.

§

What follows is an attempt to describe the internet we could have had. This could never be a simple story of cause and effect – the conditional "could have had" is essential, because it mirrors the structure of human experience: looking to the future while in the present, at the same time always wondering how we got to the present from a previous "could have been." Part I ("404 Not Found") sets the stage of this now: what was the internet we could have had? When did it disappear? What are the prevailing theories of its death and what did we think it would be? I explain that the internet we could have had was both a technically specific set of elements to which one must pay attention, and an experiment with relations among people who used it.

In Part II, "Have you tried turning it off and on again?" my arguments are intertwined not only to try to establish what went wrong, but to explain what was created and unleashed, and with which we must now live. What could have been, but also what did not happen. Each section takes us from the internet we could have had (If. . .) to the one we do (then. . .). We start with the utopias, dreams, and experiments of the internet we could have had: lofty plans to remodel the house of being, as it were. We wanted to remake our relations with one another, transform them in the name of a better, more participatory democracy, a greater freedom or more robust accountability, a better relation between markets and power.

But somehow we ended up with technologies and institutions that privilege centralized control, domination of markets, a joyless lack of autonomy, and the total evasion of responsibility. Tracing this path from one to the other requires that we note carefully the places where the details matter, if we want a bigger story about patterns and processes.

Something new was created in this "could have had" and, if we want to maintain hope, the inquiry must not just look backward but also note how these changes produced new beginnings, the outlines faintly visible – ghosts alerting us to the possibilities for power they represent. There is an AI we will have had, and the internet we could have had is inside it.

Part II contains four inquiries into the conditional: "If. . . . Then." The first section, "If Openness . . ." begins at the very heart of the internet we could have had: freedom and openness. In particular, it charts the rise and triumph of "free software" (about which I once wrote a book, called *Two Bits: The Cultural Significance of Free Software*) as a utopia of work. The section explores how open source emerged out of free software, like Hyde from Jekyll, to enable a form of arbitrary domination through openness. That openness enabled a new kind of rule, one channeled through weakly defended institutions of standardization. Google's Chromium browser is one totem of this rule.

Free software began as an ideal from the nineteenth century, ported to twentieth-century making: from William Morris to Donald Knuth. It was a utopia of dignity in work, and of the community of *Homo faber*: craft and making as the source of human freedom. Open source emphasized something different, however: the virtues of re-use, the salutary effects of transparency, and a commitment to a form of negative freedom (freedom from any interference) applied to software. Both free software and open source depended on a "hack" of the intellectual property law (new forms of copyright and copyleft licenses) to achieve this. Open source grew in popularity, in large part thanks to its essential centrality to the internet itself: the variants of the Unix/Linux operating system, the Apache webserver, programming languages like Perl, PHP, Python, and C. As its popularity grew, it buoyed the work of the largest, most successful tech companies, and slowly shifted from a utopia of workers collectively creating a common infrastructure, to a world of corporations whose employees contributed to new kinds of common infrastructures, "full stacks" and "frameworks."

By the mid-2010s, those corporations began to define the once common infrastructure according to their needs, and hence to control it. But they did not steal it: they aligned those needs with values of convenience, ease-of-access, cost, accessibility, and especially openness. A utopia of work was displaced by a utopia of not-thinking, which is a utopia we are rushing headlong into with AI today. In the process, ideals of negative liberty, of the freedom from interference as a source of autonomy, somehow enabled a form of arbitrary domination. That arbitrary domination is the source today of new questions about platforms as governers: the largest platforms today provide vital services and functions to nations around the world, and – though just barely – have started to confront their own legitimacy. They have opened up a new kind of question of popular sovereignty to which it is necessary to respond.

The second condition, "If Decentralization . . ." explains one reason why such a challenge to popular sovereignty was possible: the dream of decentralization. The internet we could have had was always going to be decentralized, it was never supposed to belong to any nation or government: not even, or especially not, the US government. This section explores the justification of the dream of decentralization according to theories of spontaneous order, associated sometimes with pragmatic concerns around net neutrality and the design of the internet, and sometimes with economic theories of Friedrich Hayek and others. Amazon Web Services is one emblem of how this dream succeeded, and then failed. Amazon's coup was to invert the internet from a decentralized one to a centralized one that could be rented from Amazon.

That coup had a paradoxical effect: it enabled an enormous explosion of innovation, open source software, and experimentation that led to the internet we do have, because people traded the dream of decentralization for the dream of scalability. The result was not a market of competing firms, but the competition of capitalisms: variants of capitalism competing to define how value would be created, extracted, and circulated by platforms. It is the source not just of a repetition of old capitalist verities, but of an authentically new problem: the problem of proliferating capitalisms, not the problem of limiting them. Combined with the arbitrary power of platforms enabled by open source, the largest platforms now use the labor of their employees – and the employees of their competitors and the volunteers who maintain these tools and the academics and legal scholars who have pioneered them – to mix and remix these structures to create something that is neither capitalism nor government but somehow both.

The third condition "If Participation . . ." looks at what has happened to one source of popular power: speech and action. Participation was central to the internet we could have had (I once wrote a book about it called *The Participant: A Century of Participation in Four Stories*). Participation is inextricably linked to speech and its powers. The internet we could have had was an experiment in tearing down all limits to speech: fighting censorship, resisting intellectual property law, circulating and leaking information, protecting privacy, fighting surveillance. In the course of doing so, we may have changed speech itself. One of the most central aspects of this fight was the simple ability to create and relate on the web: web pages, programs, data, software, new collectives. This was a new way of creating works – writing, making, coding – but it was *also* a new way of being together with other people, often very different people in far-away places with very different interests.

The combination of the technical (speech-writing) and the new relations it created made the internet we could have had participatory. It allowed anyone in, allowed anyone to speak, and this generated a new problem: Now that anyone could speak, how would anyone be heard? Who would be trusted? The Facebook Like button is one nice example of how the solution to the problem led us from the internet we could have had into the one we do have. From "the semantic web" – the utopianist's utopia – it created something strange: the *animation of deliberation*. To the qualitative magic of reading and conversation, it added the quantitative magic of polling and voting. The result is now utterly familiar: to a world of deliberation based in argument, dispute, reason, or persuasion, we have added a world of resonance and feels, vibes and virality. Whereas the former has an abundance of rules and practices honed over centuries, the animation of deliberation does not. We unleashed a new kind of speech, one we don't yet know how to evaluate, much less limit, making it a challenge to the idea of popular power itself.

The last section of Part II is a fallback case: "Else Hackers." Here I consider the case of hacking and hackers and, in particular, Anonymous. It is the "Else" case because my goal is not to ask what happened, but to consider Anonymous as the last collective of the internet we could have had and the first explorers of the one we do. The story of Anonymous is a story of the "right to know" and of the tactics of transparency, whistleblowing, leaking, and hacking. It is a story of Anonymous' speech and action, and the denial of speech and action (e.g. DDOS campaigns and tools) that emerge out of the worlds of decentralization and free software described in the previous sections. Anonymous pioneered a political version of virality, but one that was based in a logic of revelation. The rise and fall of Anonymous is a prelude to the authoritarian takeover of the internet and also of the US and many other countries as well. MAGA movements were aided by and rose directly out of the same mess of virality and anonymity as Anonymous, but they have done something much different with it: attempted to control not just speech, but meaning itself.

Part III concludes with a reflection on what the internet we could have had could have been, and how it is still with us in the "brains" of AI. Artificial Intelligence is a name that now means something quite different than what it did in the internet we could have had. Some might say: forget about the internet we could have had, we need to focus today on the threats and opportunities of AI. But this would be a mistake, because the very processes that enabled the internet we do have are also at the heart of the AI we are confronting today. The agentic and generative infrastructure being forced willy-nilly on students, businesses, governments, and artists is built out of the ideals and practices of the internet we could have had. AI is also being delivered with the same inattention to popular sovereignty that led from the internet we could have had to the one we do have, all under the cheery cover of consumer choice and economic growth, innovative promises and anti-China threats.

Was the internet we could have had a failure, or a success? It could have been both.

Syntax Error

If we want to understand the internet we could have had, we need to define our terms; there are three in this book.

The internet The internet is too big to define. I won't even try. The internet is people, the internet is networks of communication, the internet is social media, the internet is a payment platform, the internet is porn. The internet has been called "virtual" (as opposed to "real"); it has been called "disembodied" (even though it is now in our pockets and inside our brains all the time); it has been called a cyberspace (though it is awkwardly spatial). The internet is both a specific thing and a general set of effects that emerged from that specific thing. It is both a technology and a set of new relations among humans.

A 1996 US National Academies report, *The Unpredictable Certainty*, represents more or less the exact moment when a range of general terminology and theories related to a world of networks and computers, to information and communication, digitization, computerization, and the information society toppled over into a world where "the internet" refers to all of these things.[1] As the report insists, prior to this point, "the internet" was a particular kind of networking experiment: "a laboratory for the development of workable standards" and not much more. The report promises instead a "National Information Infrastructure" (NII) to come – one which we never got because the internet replaced it. And now it's too messy to define.

The internet we could have had, on the other hand, can be defined, and one can read this book primarily as an attempt to do so.

§

We What a troublesome word. We is a trap. Academics, especially those concerned with problems of history, belonging, identity, culture, or society, love to ask: "Who is this 'we' exactly?" It is an easy move: we sounds like inclusion, but with a twist of the knob, we is exclusion. We is never everyone.

But this is perhaps the reason I have to insist on it here: the "we" in the internet we could have had is a creature of an age of universalisms and globalizations. It is the outcome of a political process (very much an American one) that is precisely about expanding the franchise: a logic of perpetual inclusion until everyone is equal. A logic of speaking for others, a logic of "human rights" that look suspiciously like American rights; it is the logic of "the digital divide"; it is the logic of equality and justice before the law; even the logic of abolition democracy (though not, at the end of the day, of Indigenous sovereignty). We is not simply an ideology, but a very important kind of logic at the heart of "We The People . . .," still working through the conditional cases of the liberal representative constitutional democracies of the present.

We does not mean "humans." "Human" as a species category is irrelevant, but the "human condition" (as Arendt singled it out) is something humans do make – not a biological certainty, or an environment given to us. We therefore means specific humans appearing to each other, together in some place or another, as people who share what was made by others like them. It does not start with, but cannot escape, the we of identity (race, class, gender); it is what people try to make of being together, for a while, but then have to live with. It is not always a happy thing; many people want no part of this we. In the end, we is simply being together in its positivity, not in its exclusivity; which also means being in a space of another we or many. We is worlds, some known to each other, some not; in the best case, it is plurality.

We is popular sovereignty: it is a fiction.[2] A very important fiction because it grounds all possible attempts to limit domination and rule. For hundreds of years, and still today, that we has been the we of nations. The internet we could have had wanted a different we.

§

Could have had The conditional is intentional. It does not refer to the future but to a future in the past.

We live in a world dominated by futures: risk and risk analysis structure everything that large organizations do; techniques have proliferated for espying the future (insurance, futurology, scenario planning, etc.) or precisely managing the difference between a risky now and an uncertain future based on either a known past or a set of guesses about possibility. Most of this is organized around preventing certain futures from occurring. But what are the techniques for actively trying to make a particular future come to pass?

Academics will say that it involves imagining, or speculating about alternatives. But that is because imagination and speculation are what we are selling. Everyone else has to find a way to do something: make, build, buy, hire, promote, design, borrow, spend, argue, write, test, hustle. It is this kind of activity which makes the future; most of it is mundane, some of it will have turned out to be consequential. You can't know what will happen, exactly, when you wake up in the morning; but when you go to bed at night, the day's activity will become the stuff of dreams. It's not the other way around: you don't fulfill your dreams, your dreams tell you what you tried to do.

The linguistic form and mood of the conditional verb has built into it a logic of both possibility and retrospection: that which we *did do* but that which also *did not happen*. Sometimes we like to say "it could have been otherwise" in order to assert that possibility is always present, regardless. But it is not always safe to repeat the futures we once wanted without first learning more about the presents they have led to. The conditional is at war with causality. In programming languages, for instance, the conditional is central: it sustains the loop. "If . . . then" statements are case statements. If this happens, do this, unless this other thing happens, then do this other thing, and so on. The logic of programming assumes a finite set of possibilities, and, most of the time, predicts most of the cases. But it would not be a conditional if it could not have been something we could not predict, which is precisely where we find ourselves today.

Two conditionals in particular circulate through this book: *The Human Condition* by Hannah Arendt and *The Postmodern Condition* by Jean-François Lyotard. I explore how these books relate to the internet we could have had at the beginning of Part II.

The mid-century use of *condition* in these books is more than a coincidence; it concerns a philosophical commitment to the idea that the world humans make around them, technologically as well as institutionally, has significant consequences for the human ability to be free, to pursue justice, to create both new knowledge and engage in new forms of action through politics. Both books come from a time before the internet (and from a moment of debates about Wittgenstein, Heidegger, American Pragmatism, paradigms and incommensurabilities, violence and the definition of free speech). Because of this, I think they bring clarity to some of the issues that can otherwise seem arcane and parochial in contemporary debates about the condition we live in.

Arendt famously urged us to "think what we are doing." By this I hear her saying both that we must think and that we must pay attention to what "we are doing" with all the complexity implied by "we" and "doing" in such a phrase. It matters yet how we think: about freedom, domination, power, control, and consent, in the context of complicated technical infrastructures. Arendt worried that our ability to do amazing things had outstripped our ability to think as humans; what we have been able to do could make us dependent upon "artificial machines to do our thinking and speaking."[3] Lyotard recognized something similar: the dependence of speaking upon access to and the creation of new knowledge. Without that, we would end up with nothing but terror. Both thinkers' texts seem prescient and weirdly accurate about what has come to pass; but no such work captures things completely. There is still an urgent need to "think what we are doing" and it starts by rethinking both the history and the political theory of what we have done.

There is no question today that the internet we do have is not a condition anyone wants, other than perhaps a vanishingly small number of men in the C-suites of Big Tech. When did this happen – was it sudden, or has it always been in the making? When exactly did the internet we could have had become the one we do? The victim was last seen by hundreds of millions of people, and yet the time of death is uncertain.

- It could have been in September 1993, when the internet was invaded by the World Wide Web; no shortage of nostalgic laments were penned at the time by the denizens of the Usenet we could have had and the surfers of the C language when all of a sudden "just anyone" could connect to the servers and upload their dreck. It's an important fact that there were already people on the internet before the web arrived, just as there were already people on the web before social media arrived. Maybe each generation kills the internet it could have had.

- It could have been 1999, with the crash of the dotcom boom: the moment when the hopes and dreams of an internet we could have had were spoiled by the all-night party with venture capital and its gropy hands and slurred speech. Or maybe that was the threat of Y2K and the apocalypse that never happened? Everything is hazy. Pets.com. Beer.com. What even happened that night?
- It could have been June 29, 2007, the launch of the first iPhone, when it suddenly became clear that there would be no leaving the internet behind, at the dinner table, in bed, on the toilet. The internet we could have had was a beige desktop computer, and that is the death of desire. Apple products are smooth and white and come with no instructions. Is that the internet we do have in your pocket or . . . ?
- It could have been 2013, with the Revelations from Edward Snowden. Like all warnings of the worst to come, it came far too early and not soon enough.
- It could have been sometime around 2014, when #gamergate crept into a new era of hashtag activism, in between #blacklivesmatter and #metoo in between Occupy and the rise of the Tea Party and the volatile combination of neoliberalized demands for more diversity in tech with the unreconstructed entitlement of those in the Valley.
- It could have been 2016, with the election of Donald Trump. The first one. Or also the second one, in 2024. Was the internet we do have just the result of a political shift and not part of the cause?
- It could have been March 2020, when we kissed our togetherness goodbye and went willingly into that good night, no rage in evidence, of Zoom meetings and online schooling. The last gasp – literally – in the form of open data for Covid tracking and open source ventilator designs.

- It could have been November 2022 when, just like in September 1993, the internet was invaded by ChatGPT; no shortage of nostalgic laments have been penned in the years since, by the denizens of the internet we could have had and the high-minded defenders of human thought and writing, over the fact that all of a sudden, machines can just flood the internet with dreck. *Déjà vu* is a bitch.

Could it have been all of these, or none of them? The fact is that it seems to have happened when no one was looking, which is perhaps its own lesson, and so we are left trying to remember, to recall, to retrace the steps.

Insufficient Memory

The death of the internet we could have had has been a tricky case to solve. Even before the murder, it seems, there were suspects. Generally speaking, the culprits have been identified: Big Tech, GAFAM (Google, Apple, Facebook, Amazon, Microsoft), platforms, surveillance capitalism.[1] The internet was already garbage by 2015, and by 2018 a case was being built: weapons of math destruction and algorithms of oppression had been found at the scene.[2] Not everyone thought the internet we could have had was lost, some blamed the victim: the internet we could have had was itself too powerful, hubristic, monochrome, apolitical.[3] The death of the internet we could have had might have taken these critics by surprise.

But others said something was rotten in Silicon Valley all along: a California ideology, a new communalism, a Googlization of everything;[4] by 2011 it was already clear what the future of the internet held, and what we needed to do to stop it.[5] But somehow we were caught by surprise, and now our computers are on fire.[6]

For anyone who has worked long enough as an academic who studies what we used to call "information and communication technologies" or "new media," it can seem like a factory assembly line: every few years a new buzzword comes along: (in rough chronological order since the 1980s) cyberspace, the web, the information superhighway, the semantic web, web 2.0, social media, cloud computing, platform, big data, predictive analytics, bitcoin, algorithms, machine learning, gig economy, blockchain, web3, DAOs, AGI, LLMs, Agentic software. Academics generally tend to oblige by snapping up each item on the line and applying their conceptual or analytic tools in books and articles, starting new journals, issuing calls, coining terms.

As dreary as that work might be, it is vibrant and alive by comparison with the mainstream stories on the internet we do have. By 2025, the internet was just a boulevard endlessly wheat-pasted with stories of the Men Who Invented the Internet; the same dreadful, boring story over and over and over again, beginning always with Arpanet, touching down only long enough to recite the Names of White Men and their Tools.[7] So many of these men invented the internet that the paternity cases are still working themselves through the lower courts of the history of technology. Hark, hear the ensigns of popular history sing Alexandrian tales of Steve Jobs or Neronian ones of Elon Musk;[8] kneel down to imbibe the thin gruel of Tim Berners-Lee or Vint Cerf or Leonard Kleinrock or Marc Andreesen or any number of lesser-known men, and let them harangue you with stories of what went wrong with the internet they invented – because, like absent fathers, they now find someone else to blame.

The internet we could have had promised other stories and produced them. Joy Rankin and Kevin Driscoll and Charleton McIlwain have other stories to tell, more interesting ones.[9] Lisa Nakamura and Wendy Chun have long resisted the colorless and disembodied theories of the internet. Marisa Duarte brought to light the work of an indigenous internet we could have had. Ruja Benjamin, Katherine McKittrick, Alondra Nelson, and many others detail the work of Black communities, from technologists to Afro-futurists. Avery Dame-Griff has revealed a part of the very long history of the transgender internet, and Daniel Greene has revealed class and property stories that ground the internet we could have had. So much more has been researched and written.[10]

But you won't find these stories on the internet we do have, only on the internet we could have had (and this is part of the problem). On the internet we do have, scholarship that contests or complicates things disappears from view in favor of the stories most likely to be read, and so therefore those that are re-written over and over again. Worse still, the scholarly machine itself succumbs to the demands of the internet we do have; it entices us to abandon the internet we could have had. Academics mistake this process, or assent to it, in an attempt to deepen the conversation in the public sphere; we rush to contribute to it by writing "trade books" arguing with stories we don't want to tell and re-tell. But the approbation of audience numbers and visibility, which is now structured by the internet we do have, forces our hand. We are trapped.[11]

We are trapped in the sediment of stories of heroes and invention – or their opposites, greed and villainy – even in the most sophisticated accounts. Even works most critical of the internet we do have, zealous in their targeting of capitalism, complacently repeat the stories of invention rather than risking a more complicated story, or unearthing an untold one. They blame capitalists instead of capitalism, yet the suspects are all too usual – and all too often reduced to a single place, Silicon Valley. A filo of stories, both hagiographic and critical, has made a regional shorthand into a thick blanket of simplicities.[12] Everything bad happened in Silicon Valley, birthplace of "cyberlibertarianism," from the proto-bro garage heroes of homebrew computing to the creators of a gig-economy, an asset-stripping private equity fire sale, a real-estate and environmental catastrophe, and perhaps most darkly of all, a new form of capitalist extraction not of land, labor, or rent, but of the very soul of human beings themselves. Yet even in Silicon Valley, there are other stories to tell.[13]

In the age of the "techlash," stories of our technological inheritance culminate ultimately in grand theories of their own, often dispensing entirely with "the internet" and telling instead how Big Tech has become nothing less than a confirmation of human evil, or human stupidity, or both. Let's consider them in turn.

§

Shoshana Zuboff has been watching these disasters unfold for decades, and has written consistently on the failure of the information, computer, and internet revolutions to yield any of their utopian hopes. Starting in 1988, Zuboff wrote an eclogue of computerized labor, *In the Age of the Smart Machine.* At its outset, it was to be a chronicle of a major transition in everyday life on par with the industrial revolutions of the eighteenth and nineteenth centuries. But in retrospect, it was also clearly a warning bell about the horrors to come; in it she built a careful lyric of the fields first tilled by hunched secretaries, eyes sore with cathode ray tube (CRT) screen glow and arthritic data-entry fingers. All these bodies and all this labor failed to solve the "productivity paradox" of automatic office work, and instead introduced a raft of new ailments.[14]

By 2019, Zuboff's passion for the sweeping narrative had not slowed; in *The Age of Surveillance Capitalism* she was able to write "Just a moment ago, it still seemed reasonable to focus our concerns on the challenges of an information workplace or an information society. Now the oldest questions must be addressed to the widest possible frame, which is best defined as 'civilization' or, more specifically, *information civilization.*"[15]

Zuboff's account of the arrival of surveillance capitalism is solid. At 600 pages, how could it not be? It is a story of "instrumentarian power" on the rise, and the details she relies upon are some of the same ones I will point to at the moment of transition between the internet we could have had and the one we do; especially those having to do with how "behavior" is transformed into automated and algorithmic processes. She is unafraid to call out academics like Hal Varian at Berkeley and Alex Pentland at MIT who have played willing court counselors to the oligarchs of Big Tech, and the book is an avowed updating of Arendt's perceptive understandings of totalitarianism and power.

But it is also a book that, if you squint, is indistinguishable from *The Matrix* – one of the gospels of the internet we could have had. Zuboff's story is of a capitalism of perfect control, perfect knowledge, in which humans are both labor and raw material, completely encased in glass containers, perfectly visible to the capitalists. The book uses the image of the "hive" repeatedly, in telling a story of the systematic accumulation not of labor or land, but of behavioral surplus. The consequences of this are the "loss of home" or what she seems to suggest is the same, the "loss of humanity." As an aside, it is also an anti-Chinese story as well: China repeatedly plays the role of a society of perfect surveillance and missing freedom, the destiny of a US under surveillance capitalism. At its feverish pinnacle, the risk is that the oligarchy will have taken over control of *popular sovereignty* itself: a fundamental evacuation of human power and subjectivity from the world.

This is, to state what should be obvious, a politically disabling conclusion. It is a conclusion dangerously close to the very nihilism often on display among the most extreme characters of the Silicon Valley pantheon, the Musks and Thiels and Yudkowskys. All Zuboff is left with is a nostalgia for the time of Adam Smith or Henry Ford, a time when capitalism had a "social contract" of reciprocity with consumers and laborers, limited by law and yet productive of value; or when promises mattered because they were made by humans, and not algorithmically executed by machines: "The convergence of freedom and knowledge transforms surveillance capitalists into society's self-appointed masters. From their high perch in the division of learning, a privileged priesthood of 'tuners' rules the connected hive, cultivating it as a source of continuous raw-material supply."[16] The only solution is an extreme one – to incarcerate them: "Surveillance capitalists know too much to qualify for freedom."[17] The internet is a result of human evil gone unchecked.

As a clarion call, I agree: I would hate to be on the wrong side of history if Big Tech turns out to be worse than the Third Reich. It is not, however, out of school to suggest that the Matrix-style control she descries is in fact otherwise. First, it is never complete, as domination never is; what she describes relies on the willing participation of both employees and consumers at these firms, and in ways that are not subject to behavioral control as she describes it, but rather an honest embrace of specific desires. Second, it may be that the design of this system of behavioral extraction is the result of popular sovereignty itself, not of a coordinated coup. Which is to say: it is embraced by diverse actors beyond the green campuses of Silicon Valley giants. If that is the case, it is because of the existence of freedom of a particular kind, granted to a much larger range of people, which enables different forms of engagement with our world, and whose only check is that there remain countervailing forms of freedom. The internet we could have had was built on this promise, and yet we got the world that Zuboff describes. The puzzle is unsolved, but the killer is not Silicon Valley.

§

Opposite this story of evil is one of stupidity. Marion Fourcade and Kieran Healy are willing to blame a much wider set of people for the disaster of the internet we do have: users. In *The Ordinal Society*, they explain how users consented to the disaster because it was "delightful" along the way. As in Zuboff, this book is focused on explaining how the internet we do have was always the one we were going to get, if only we had been smart enough to see it. Whether by outright manipulation or just to please consumers, Big Tech grew large and powerful because people enjoy convenience, the "affirmative pleasures of connectedness," and the "magic" of technology. We make the same bad decision over and over again; we are addicts. But more than that, we are hard-wired by status: we will ruthlessly compare ourselves to each other, and the more tools we have to do so, the more we will do so, tightening the net of Big Tech.

Their story is compelling: ranking and ordering both objects and people is an activity of widespread practical use, whether it is restaurant reviews or hiring decisions. For centuries such forms of ranking and ordering have been features of solidarity within societies, formalized in institutions, and seen as reflective of social imaginations generally. Big Tech has changed all this by providing widespread tools that bypass or erode these older forms of solidarity. They "exfiltrate human sociality by means of the gift"; they provide simple and easy-to-use tools of "quantified and ranked comparison"; they produce an "aesthetic of justified hierarchy" throughout our social lives that we "no longer control." People consent to it because, being social beings, it sort of feels right to be able to make such distinctions; but it "blinds us to what we all share, and chips away at solidarity in feelings and institutions."[18] So far, so good.

But users are not the only dupes in this story. *The Ordinal Society* doubles down on the idea that the internet we could have had was nothing other than a delusion of utopian idealism and irrational inevitability, most clearly represented by the ideologues of free software and open source in the 1990s and 2000s. To do this, they must willfully misrepresent both the history and the subsequent scholarly work on the topic.[19] But it's not clear whether they build much of a case for it in the end: they simply abandon any belief that there was ever an internet we could have had.

They might have picked a different target for this purpose, not open source but Karl Popper's *The Open Society and Its Enemies*. Long before the ideologies of open source, open access, open science, and so on, Popper was hailing modernity as the best of all possible worlds. Popper defined openness as a necessary feature of governing modern "abstract" societies: the limitless, extendable, evolutionary horizon of knowledge for governing. Popper reworked the history of philosophy through the concept of the open society. Openness lay opposite tribalism and tradition because it invited constant innovation, progress, and experiment. The opposite would be stasis, repetition, and a dream of perfect pure control without change, which Popper ultimately accused Plato, Hegel, and Marx of perpetrating as "oracular philosophies" immune to the use of reason.[20] Abstract societies consist of individuals, and individuals confronted by the necessity of personal choice (and status!) rather than the slavish following of magical or traditional rules; they are distant from one another and in need of super-human forms of coordination.[21]

The point of recalling Popper here is that contemporary societies are in fact complex, reticulate, "abstract" in many ways, and governed by knowledge of diverse kinds embodied in increasingly diverse institutions with complicated governance structures. The "rules" of the game are not simply obvious to everyone. But it is Popper's appeal to the need of abstract societies to rely on things like creativity (or creative destruction), innovation, the growth of knowledge, and the fundamental openness to change, which is background to the kinds of explanations that *The Ordinal Society* provides. One wonders who would be the enemies of the Ordinal Society if their title mirrored Popper's.

By blaming people like Richard Stallman or John Perry Barlow (such low-hanging fruit!), Fourcade and Healy's book sacrifices the very heart of their story: that diverse, polycentric, distributed collectives of people – and not a handful of Big

Tech corporations, nor a set of bearded ideologues of software freedom – created the tools of this ordinal society. *The Ordinal Society* itself is fully aware of aspects of this story, and wants to reveal for us how that "mundane" data (logs of our actions, clicks, and Likes) came to be much more valuable and meaningful than the data we associate with a more august form of privacy. But it comes to exist not simply because it is valuable to corporations, it has inherent interest for all sorts of people: for social scientists, consultants, middle managers, and marketing grunts as much as capitalists or users.

Many people, it turns out, want a smidgen of surveillance and behavioral control in the service of making things better in the world around them. They don't simply "delight" in the tools provided to them by Big Tech, they actively participate in making them in the first place. The doctor and the nurse want to change the health behaviors of unhealthy people and so thrive on better data, organized more clearly, in order to do so; the HR middle manager wants demographic data for improving diversity, equity, and inclusion (RIP!); the education researcher wants to demonstrate the improved behaviors resulting from a different way of teaching or testing and so welcomes the data on children in schools; the activist wants to change the narrative, and so needs to invest in memes and imagery and the tools for generating engagement. "Surveillance capitalism" might better be labeled "surveillance liberalism." In both cases, it is more than surveillance, it is a way of exercising forms of control and governance, sometimes clothed as convenience, sometimes as paternalism, most often as freedom.

The internet we could have had did in fact put a dangerous tool in the hands of people: the ability to remake relations among people by animating those relations through software and computation. People all over the world have used this tool to envision a collective, but one better than the one we had: a collective that was intended to get free from exactly the structures and processes, so ossified in states and corporations of the twentieth century, that seemed to support both human evil and stupidity.

§

Ironically there is one assessment of the internet we could have had that is eminently more accurate, more fun to read and more hopeful than any other, but it comes from someone whom Healy and Fourcade might consider one of the most accomplished dupes of all: Cory Doctorow and his theory of enshittification. His diagnosis of the changing landscape of anti-trust law and policing, the gutting of the power of self-help (by which he means interoperability of devices and things like the right to switch or exit from a technology), and the precaritization of labor are rich veins from a life's work of advocating for the internet we could have had. He has a passion for saving it because he fully grasps the way that, quoting William Gibson, the internet has "everted"; it is no longer a cyberspace in or out there, but a logistical system that controls and directs the world of physical goods, from the cars we drive ("computers we put our bodies inside") to the groceries we buy (automatic digital pricing on store shelves) to the jobs we may or may not get and hold on to. If you need a political platform – which is to say, if you wanted a book about "how to save the internet we could have had" – then Doctorow's is the one to embrace.

What is missing (from a political theory perspective and thus perhaps less urgent than the practical action we might need right now) is something like a sensitivity to historical consciousness, or the haziness of the future. When the internet was new, it was not at all clear what it would be, and we fool ourselves now if we insist that we knew then what it could have been. Choices are like that, or they wouldn't be real choices. The deeper question is whether the future has changed; and whether the reason the future has changed is because of our attempts to make a different one. Our relation to the conditional mood has shifted; progress is no longer what it was in the nineties – if for no other reason (and there, are a few: knowledge of climate change, the Anthropocene, irreparable income inequality) than the fact that the future seems a lot shorter and less certain to many people than it once did.

Something certainly went wrong with the internet we could have had, but it was not totalitarianism, and it was not a delusion. It was a beginning. It was the creation of the internet we could have had, which set the terms for the internet we do have. Given this, it is not necessarily clear (or any clearer than it has ever been throughout history) where the "surveillance capitalists" came from, or how they gained their power; nor is it necessarily clear that anyone on the internet we could have had was simply duped by utopian dreams or subject to a vast, unprecedented bout of misrecognition. Perhaps it is simply that "the means used to achieve political goals are more often than not of greater relevance to the future world than the intended goals."[22]

Reloading . . .

What, then, is *the internet we could have had*?

The internet we could have had is, in a narrow technical sense, still the one we do have today. For anyone who spends a lot of time configuring and deploying servers or web applications, struggling with DNS or digging into protocols and RFCs, the core technologies have actually changed remarkably little. Linux and Apache are now 30 years old; the core protocols of the internet are close to 50. Artificial Intelligence runs on Python – one of the earliest languages to emerge on the internet – and every fancy new stack or framework of the month builds on top of these older technologies. The most consequential US laws governing the internet include section 230 of the Communications Decency Act, passed in 1996, and the Digital Millenium Copyright Act of 1998.

And yet it is also very different. The internet we do have is a centralized cloud and data-center system, the pipes and data are run by very large companies with very sophisticated engineering and monitoring capabilities that keep the internet on, all the time, and fast. Content Delivery Networks determine the shape and speed of the internet, as well as the global location of content both mundane and sensitive; access has shifted from modems and ethernet to cell-phone networks, ubiquitous wireless routers and Elon Musk's Starlink, which has lit up nearly every spot on the globe – good luck escaping the signal. Today there is the General Data Protection Rule of the EU, and now we are "data subjects."

The internet we could have had is both a changing technical infrastructure and an evolving process of bringing people into relation with each other. It has to be understood as both, and can't be reduced to one or the other. It is not an accident that a parade of theoretical terms has tried to suture this either/or: network society, virtual community, digital markets, new economy, republic 2.0, and worse. In this section, I want to paint a picture of how the internet can be both things, how we have to think about it being both things if we want to make any theoretical headway at all.

§

For many people the internet we could have had is a technical thing first. It is computers, software, networking, but especially code, protocols, formats, interfaces, frameworks, stacks, and standards.[1] It starts with these things, but is more than the sum of its parts. It is exciting because it is emergent and complex. It is exciting because it is distributed, decentralized, and under no one's power, but a result of everyone's actions. It results from actions presumed to be, if not good, then at least self-determined, and therefore more democratic, more open to participation.

Narrow technical definitions are partially helpful: the internet is a suite of protocols (TCP/IP, UDP, email, HTTP, etc.); it is a "distributed computer" (a networked system made up of many computers, able to access each others' services); it is a material infrastructure (cables, routers, waves, light, switches, and specialized machinery to keep it all running); it is real estate (data centers, carrier exchanges, server farms); it is a theory of communication structure (packet switching, robustness against attacks, end to end). The list could go on.

Any one of these definitions, or even a collection which bundles them all together, will be inadequate. It leads all too easily to thinking of the internet as either unchanging, or as a technology whose precise parts and mechanisms can be accurately and uncontroversially identified and innovated.

The internet is always changing, and in more than one sense.

First, as a technology, it was designed and implemented to be open to certain kinds of change, and this is what made it exciting and novel for a lot of people who engaged with it. But, in a second sense, it was also subject to forms of change that were not part of its design or implementation: changes in how people changed it, so to speak, because of how it was perceived and valued, as open, flexible, extensible, and modifiable.

Both forms are "technical." The internet is a technological object that changes more or less autonomously, responsively, or mechanistically: packets are routed, new nodes come and go, files are updated, software sorts and calculates at the speed of light. But the internet is also a technical object changed by human hands, coded and recoded, updated and reconfigured, legally constrained, extended as part of an effort to respond to fiscal or economic constraints. Individuals, but also groups of people, can plan and execute changes.

Both forms might also be "democratic" in a certain sense: an imagination of the capacity – and the right, the entitlement even – for anyone to engage in solving the problems that confront them by changing the internet directly. It is not a democracy grounded in existing representative institutions. Changing the internet we could have had was not something that had to be done only through the activities of election, legislation, judicial oversight, or through the popular practices of protest, speech, association, or advocacy. It was, rather, something that people could just do – by banding together and making the internet into something imagined to be a solution to a problem.[2] Such banding together need not be commercial, but, for many people, nor did it restrict commercial activity.

The internet we could have had was a government but not government. It was a particular version of popular sovereignty – a pragmatic democratic freedom enabled to create new and different institutions: self-organized, polycentric actions begun through direct popular sovereignty. The schizophrenia of this approach was obvious. It was the creation of a new state but also the struggle to figure out how the internet we could have had articulated with the existing institutions. Create a new constitution, or amend the one we do have? Pragmatic approaches tinkered in isolation from the state precisely because the internet we could have had was a "virtual" reality, a second life, a noosphere (or a *terra nullius* beloved of colonizers and homesteaders).[3] It was possible to imagine a kind of do-over: what didn't work the first time could be tried out differently, with more power and a technology suited to the purpose.

There are key differences between these two forms of change: on the one hand, a changing and changeable machine – designed to be constantly evolving, and also open to modification and extension; on the other, a machine made of people, consciously being amended and transformed by diverse groups and individuals. The two forms are different, but they necessarily became entangled with each other.

§

A changing and changeable machine There is a *resonant contradiction* at the heart of the internet we could have had. A contradiction because it is a precise, exquisitely engineered structure which will nonetheless produce an unpredictable, but spontaneous, order. Resonant because even this technical object is not precise; it rhymes with other ideas, but allows difference. The internet we could have had was a technology, but it was also a space of endless possibility that could have been, at the same time, a permanent solution to the thorniest problems of the present. These contradictions formed the vital substance at the heart of the internet we could have had.

One could claim – and many do – that the internet we could have had simply emerged from the design of its protocols. This does not necessarily mean that such a design determines the outcomes; rather, it is similar to how a political order arises from its constitution, or a corporation from its charter. David Clark, a key designer of internet network protocols and structures, and one of the most sophisticated direct participants in its history and evolution, has long made the case that the design choices of the core suite of internet protocols determined the shape of the internet we do have in this manner. For Clark there is no "could have had," there is only the internet that unfolds as a result of these concrete and technical design choices.[4]

If the internet were fully determined by its design goals, however, its outcome would be unsurprising – there would be no internet we could have had: we would have the internet we designed, perhaps with some "unintended consequences." But Clark's claim about the internet is one of determinism and unpredictability at the same time (recall the name of the 1996 National Academies report, to which Clark contributed: *The Unpredictable Certainty*). In this story, the internet is not an organization, a market, or a hierarchy, or even a network. It is a network of networks, a scheme, a constitution even, for connecting up networks. It is not the network itself, but the rules for a constantly changing network. Clark has long been concerned to show us that a set of choices about how the technology works – even a changing, dynamic, reconfigurable network – constrain the structure and function of that technology; this was, more or less, intentional on the part of the designers. Certain kinds of problems or limits, if not the exact form of the internet we do have today, might have been predicted from the initial design of its protocols.

Internet engineers like Clark talked about the genius of the design of the internet in terms of its packet-switched networks and the ability to extend or reconfigure the topology. But many other people, encountering the structure and design of the internet for the first time, did not see strict rules creating something unpredictable, but something that was endlessly malleable yet permanent. Rather than a limiting, but unpredictable structure emerging from a constitution, many people saw the internet we could have had as a fundamentally unlimited and liberating technology, a space of *endless possibility.*

That endless possibility came in different variants; there was the cyberpunk-fueled possibilities of *Wired* magazine and *Mondo 2000*; there were the temporary autonomous zones, the netwars, the anti-globalization networks; there were cypherpunks and anarchists.

Imaginations of the internet we could have had floated among an archipelago of abstract dreams of endless possibility in the sciences of the 1990s: complexity, indeterminacy, chaos, and non-linearity. These ideas saturated 1990s engineering research; books about chaos and complexity lined the popular science shelves of the bookstores we once had.[5] These abstractions took rigorous mathematical forms as well as culturally resonant ones. *Gödel, Escher, Bach* was the *On the Road* of the late eighties and early nineties; Prigogine and Stengers' *Order out of Chaos* translated physical non-linearity and chemical far-from-equilibrium chaos into philosophies of becoming and cosmopolitics.

In such a soup, the internet we could have had, which could be approached mathematically because it was grounded in software and hardware, was a toy model of universal chaos and complexity. Local dynamics could exhibit negentropy, building fractal relations across thresholds and boundary conditions. In a network of networks, the internal operation of "administratively bounded networks" could be ignored and left to their administrators, so long as they interoperated fully with the internetworking protocols to produce an order beyond any overseer's power. The resulting internetwork was therefore constitutively unpredictable and always changing. It was, in short, alive; it was a proto-ecology.[6]

As such, the endless possibility could be recursively applied at every layer of the technology. There were no "fundamental rules" and yet there was nonetheless a pure possibility, something to pin all manner of hopes on.

The exuberance of endless possibility was different from the engineer's protocol. In a story like Clark's, the design sets the initial conditions, the final state is indeterminate, but the paths and possibilities are constrained by these technical design choices; the goals that are set in advance determine the paths by which the internet becomes what it is, even if those paths are unpredictable. But the story of endless possibility is a frankly more theological one: a story not unlike that told by contemporary Artificial Intelligence stans, in which chaos and complexity give rise to novel, emergent properties, and maybe thus to consciousness and intelligence. Long before the AI of today, we had the "noosphere" of the 1990s and an efflorescent "artificial life."

However, even in Clark's version, the internet we could have had did not come to pass quite the way he suggests it might have. In a famous paper, Clark enumerated the eight goals, ordered hierarchically, that the designers of the original internet attempted to achieve at a technical level.[7] These included things like allowing different kinds of communication services, emphasizing survivability, privileging the ease of joining the network. Above all, it privileged a notion of the "datagram": the packet, or the connectionless protocol, designed to allow information to get from any point on the network to any other point, by any possible route, also known as the "end to end" principle (also known as the "surviving nuclear attack" narrative of the internet's origins). The initial design did not include goals like security and encryption, or privacy of user data, nor did it (explicitly) privilege any particular institution as one of the "ends."

Over time, the ordering of the stated goals, and the lack of the unstated ones, became more apparent to people. This was not simply because of the property of the protocols, or the ordering of the goals, but because people increasingly used the internet for different purposes. The world around the internet had changed – one might say dramatically – from 1973 to 1989 and beyond, and the needs and possibilities people brought to it did as well. Contingent though these choices were, they were real technical challenges that had to be figured out, and once figured out, set the stage for others to encounter the internet we could have had as it was designed – and then re-designed. Re-engineering the internet for security, for instance, has turned out to be one of the more expensive and longstanding challenges to the initial design.

The *very openness of the system meant that people could break it.* Instead of aiming at the design goals, they were enabled to find ways to pursue other goals, and in the process created new problems. Thus a tension between the internet we could have had as a set of constrained design choices with unpredictable outcomes, and the idea of it as a space of infinite possibility.

Clark makes the conflict between unpredictability and infinite possibility clear in the case of the entities that argued for disrupting the "end to end" design of the internet: governments who insist on their right to spy, or ISPs who claimed a need to manage traffic loads, or were being asked to spy on their own customers for various reasons.[8] Such uses were changes in both of the senses above: to change the way the internet worked by asserting a right to change the way it worked. The paradox was clear in other cases too; new uses, different incentives, creative discoveries all resulted from a widening engagement with the internet itself, which changed what the internet could do.

Because it was "democratic," people were enabled to use this technology in dramatically different ways, because this flexibility was a core goal of the internet we could have had. The fact that there was no coordination among such groups meant that this would almost inevitably become a contest of an unusual sort. It was democratic popular sovereignty without any form of coordination across a radically distributed and plural "we" of the internet we could have had.

In making the internet we could have had, the unpredictability of the internet did not have limits. In fact, it was experienced not as unpredictability or indeterminism, but instead as endless possibility: a technology whose goals were not set in advance, a technology that could become or do anything. The result would be an internet that no could no longer tolerate the very activity that brought it into being: democracy.

§

A machine made of people No matter how often storytellers reduced the internet to a technology – to its protocols and standards and markup languages and software – many other people told a different story: the internet we could have had was made of people.[9] It was made of real people interacting in ways never possible before; not programmed computers interacting, whether deterministically or not, in some autonomous way, but humans being together with the help of such tools. The internet we could have had was a space in which human beings experimented upon their own relations. And in this context, the "technical" internet (specific protocols and software and engineering designs) rests alongside dozens of "draft versions" of the internet we could have had emerging in this moment. What links them together are these experiments with relations.

The technical details matter, of course: PLATO was not the Internet, FidoNet was not the internet, but in both cases there were elements of experimenting with relations that occurred across them, and were sometimes ported from one to another. Some of the earliest experiments with time-sharing in the MULTICS project set paths of interaction: usernames, passwords, pseudonyms, ways of seeing who was online and who was idle. The brief late 1960s business fad for the "computer utility" (cloud computing *avant la lettre*) elevated "multi-user" computers into something more like a grid or infrastructure. The PLATO project or the rise of the Bulletin Board Services or Minitel, from the 1960s to the early 1990s to FidoNet and Usenet – networks often very different in technical design from the internet, but possessing some of the same experimental forms of collectivity and interaction.[10] What opened up for exploration was not a controlled experiment in technical networking, but a live world: people interacting via networks and discovering thereby new ways of interacting. New relations became possible: not a model of relations to be tested, but an implementation of such relations themselves, a model that people could put themselves inside.

The medium of all these draft versions of the internet we could have had was, for a long time, a kind of speech-writing, mundane and momentous. It was like speech because it was immediate; it was writing because it was permanent. Precisely because it was an emergent and changing thing, speech-writing on the internet we could have had messed with finality: was something written on the internet written forever? Was each post a permanent work, or was it part of a conversation, ephemeral, emotionally laden, purposeful but amendable? What of this would be kept and what would not? Who would keep it: letters in a drawer, or books in a library or something else?

Speech-writing was an experiment: performative speech acts in a more-than-linguistic sense, precisely because it extended to software code, protocols, legal licenses, documents of all sorts that could be animated, or could have technical outcomes. We created "disturbingly lively machines."[11] People were testing out relations that quickly became more than tests, but relations filled with both love and violence, carried out in something like writing.

Speech-writing allowed novel forms of doing by saying. Speech-writing did not simply create or circulate information, it was a way of forming relations by saying them, or writing them, into existence. The temporality of this interaction was not quite the immediacy of embodied speech, but it was far from the imaginative postal slowness of the epistolary romance. Elements of both rendered visible the space in between them, for people to explore.

The internet we could have had (and its draft versions) was thus a kind of telescope: a telescope of being together; a tool for bringing close, taking apart, and then remaking relations with other people. In Arendt's story, Galileo's telescope changed the human condition: "it put within grasp of an earth-bound creature and its body-bound senses what had seemed forever beyond his reach."[12] Unlike that telescope, the internet we could have had did not present a distant physical reality – bring the impossibly far close to perception – it rendered visible *relations*.[13] The material of the internet we could have had was not sense-perceptions, not the visual or the auditory, not just the text-powered imagination of the novel; it was above all the relations with others that suddenly appeared, as it were, up close, when before they were distant. They were not relations of any fully known kind: they were similar to, but different from, the ones people thought that they had known – familiar, social, economic, communal, urban, formal, informal, hierarchical, anarchic, at a scale and with a form that had never been experienced before.

The telescope of being together not only let us see something we could not perceive, it allowed us, as computational technologies do generally, to *animate* something. We could set forth – write – some rules and data, and watch as a machine calculates with them at a speed immeasurably faster than humans can. We could take a model of relations, from the simple interaction of two or more people to the complex relations of a community, a society, a market, and then use the tools of the internet to animate those relations using software and networking technologies. We could imagine a model of an economy (supply and demand curves, some game theory, or an econometric relation) and collect some data (people buying and selling magic daggers and breastplates in a dungeon, tallied in a database) and animate these things to form an emergent set of relations like an economy, but both different and susceptible to repetition and modification. Not a model, but a simulation; not just similar, but possibly better.

The relations implemented on the internet we could have had started with theories, axioms, patterns, ideas, and ideologies from different histories and traditions. But the goal was not to repeat them, the goal was to make them better. Better because every understanding of these past relations also came with a critical sense that they failed to live up to ideals, but could be improved by the application of new technologies perceived as powerful and transformative and inevitable. Markets, for instance, suffered from externalities, failures, and inefficiencies; societies suffered from inequalities; organizations suffered from bureaucratic sclerosis or domination; social movements were limited by time, place, and manner; communities suffered from entrenched biases or hierarchies of racial control. They could all be improved, and the internet we could have had was a place to try.

These were not abstract and fuzzy ideas. Elinor Ostrom's Nobel prize-winning work on commons was a critique of ideas of what a market could do and how it could fail.[14] Dimaggio and Powell's iron cages of institutional isomorphism guided thought about the re-design of organizations (alongside a raft of business literature gurus like Peter Drucker or futurists like Alvin Toffler); Manuel Castell's theorizations of networked social relations was grounded in critiques of changing urban forms and their possibilities for social movements.[15] Hardt and Negri consecrated a new collective – multitudes – and sketched a new empire that galvanized anti-globalization protests and gave nascent techno-activist momentum to the Occupy movement; Walzer's communitarian critique of liberal individualism raised a challenge to neoliberalism's demand for individual responsibility.[16] Political theorists like Joshua Cohen and Amy Guttman elaborated variations of deliberative democracy as a critique of the narrow election-focused party politics afflicting our democracies.[17] Performance artists like Brenda Laurel or Guillermo Gómez-Peña created novel aesthetic and performance practices and political counter-movements.[18] Business school scholars like Clayton Christensen elaborated theories of disintermediation and the innovator's dilemma.[19] Science and Technology studies scholars such as Latour and Haraway broke down the opposition of science and society and even the line between human and non-human to insist on relations, networks, and associations.[20] All these debates and more were vital to the people figuring out the internet we could have had, some more explicitly so than others.

Although diverse in sources, many of the experiments with relations on the internet we could have had maintained and amplified a bias toward universality.[21] The key technologies involved – general-purpose computers and networks – were understood to be universal, which is to say, useful for and applicable to any possible domain or way of living: economic, social, private, religious, or aesthetic. Universality also meant global, planetary, transcending the existing limits of the collective kinds that entranced people in the centuries leading up to it: nations, societies, colonies, and communities. Computers were already recognized as intervening universally in domains where super-human forms of calculation and organized control could be imagined; but adding generalized computers to generalized networks in the internet produced the novel sense that this tool could transform more than just the daily activities of specific forms of work or leisure; it held out the possibility of changing relations among people generally.[22]

What most puzzled observers (especially journalistic accounts) about the animation of relations on the internet we could have had was that it was not just a simulation or a game or a theatre; people really began to live inside these models, for real. They started to really buy magical daggers, for real money. Explanations at the time had few repertoires to draw on to understand what had been made possible. Many resorted to platitudes of real and unreal: virtual reality, second life, role-playing. Whole books about the internet as theatre could be and were written.[23]

To say that the internet was a "virtual world" was not incorrect. From the earliest text adventures through MUDs (Multi-User Dimensions/Dungeons) and MOOs (MUDs, Object-Oriented) on to *Second Life*, at first these things could only be called a *fantasy* separate from the real world.[24] A great deal of distributed enthusiastic creativity centered on the idea that these new relations could take place disembedded from a physical world that some people felt trapped in: on the internet no one knows you are a dog. But the human worlds were no less real for that. Worlds of relating: being together in Arendt's sense, not worlds of perception or vision or light in Galileo's sense. It was apparent early on that the people engaged in these draft versions were working on new kinds of selves, new relationships, new subjectivities, new collectives.[25]

Experiments with relations on the draft versions of the internet we could have had were all too real; not a world of perception and imagination but a world of relations with other people, even if avatars, pseudonymity, "no one knows you're a dog" were the norm. There was (maybe even until Edward Snowden's revelations) an intransigent cultural and political resistance to the idea that what happened "online" was real. Despite the increasing spread, as more and more people joined, of these relations from six to five to four to one degree of separation, few people thought they would seriously matter to the way things would go in the world. The cognitive dissonance that defines the turn of the millennium was this: the immense demand and desire to buy into the technical hype coupled with the profound refusal to believe that it would be consequential – that it would really, honestly, and permanently, change things. For many, especially those who were not really on the internet, the internet we could have had did not matter to the realities encasing or supporting it. It could be – could have been – nothing more than a phase, a game, or a distraction from more serious, more real, endeavors. Until suddenly it was reality itself: the internet we do have.

II Have You Tried Turning It Off and On Again?

> The *polis*, properly speaking, is not the city-state in its physical location; it is the organization of the people as it arises out of acting and speaking together, and its true space lies between people living together for this purpose, no matter where they happen to be . . . It is the space of appearance in the widest sense of the word, namely, the space where I appear to others as others appear to me, where men exist not merely like other living or inanimate things but make their appearance explicitly. This space does not always exist, and although all men are capable of deed and word, most of them – like the slave, the foreigner, and the barbarian in antiquity, like the laborer or craftsman prior to the modern age, the jobholder or businessman in our world – do not live in it. No man, moreover, can live in it all the time. To be deprived of it means to be deprived of reality, which, humanly and politically speaking, is the same as appearance.[1]

The internet we could have had was, for a while, the *space of appearance*. And then it was gone, reality disappeared. The internet we could have had was also, for a while, the home of GeoCities, a now beloved relic of the 1990s internet. The marriage of high and low is deliberate here: the internet we could have had is serious business and concerns the very capacity for popular political power; it is also a story about some forgettable, sometimes regrettable, joyful and ugly, technical choices.

GeoCities was a city, but Geo. Urban relations, but in a new kind of neighborhood: animated gifs and web-rings, comic sans fonts and blurry pre-Instagram personal photos, experiments at the very origin of web design and traces of communities and organizations and events and companies and projects, some that never got off the ground and others that went elsewhere. The rush of creating a GeoCities site was not technophilia, it was the exploration of something like a new neighborhood, with new neighbors, shops, bars, criminals, and politicians, inside which one could live, differently, maybe better than one could in one's own physical urban neighborhood.

This was a space of appearance: a place where people appeared together, to each other, and engaged in forms of speech and action whose consequences then unfolded in ways they could not predict. It was not the only such space of appearance on the internet we could have had, but it was a good one.

In 1958, long before GeoCities, Hannah Arendt wrote *The Human Condition*, her most concentrated statement of her theory of politics. Arendt's theory rests in her core disagreement with the very long tradition of political thought: power isn't about domination, it is a result of individuals engaging in speech and action in the space of appearance, and the things they begin by doing so. Almost all other traditions in political theory associate power with rule and authority, from Plato to Hobbes to Rousseau to Rawls. Arendt tried to separate them. She insisted that power rests in people coming together in a space of appearance where they set things in motion; all else is the appropriation of, or restrictions on, that unfolding process. Violence and authority are attempts to appropriate power, not to express it. Popular sovereignty – of the sort we associate with liberal representative democracies in the West – was for Arendt a result of the kind of power she described, not its source.

Arendt's vision is that "speech and action" are distinctive expressions of power; they are different from both strength (an accumulation or possession that can be directed to different ends) and violence (the mute brutality of forcing others to do what one wants them to do).

Arendt offered, for instance, a definition of power that is not about domination or rule (i.e. not about it being a possession). Her definition of the relation of speech and action is worth pondering. Action is "finding the right words at the right moment, quite apart from the information or communication they may convey."[2] Action can only happen if a speaker *and hearers* are together at the right moment; and if they are it will have an effect, it will *begin* something. The agora of ancient Greece is her model for thinking about how this might happen; such small-scale versions of democratic action still re-occur in times and places throughout history. But the agora is not essential to this definition. It challenges us to think beyond the particular ways in which we communicate, how information is processed, what media are used, or with what kind of people, in what kind of places, this all takes place. The raw fact of this definition is a challenge to all theories of power that are intent upon taking it away from the people who are together at the right time, and find the right words, to begin something. That, for Arendt, is power. The rest is what we – or they – do with it.

If the internet we could have had is an experiment in animating different kinds of relations, then it poses a kind of trouble for Arendt's definition of what it means for humans to appear to each other, to be together: "To live together in the world means essentially that a world of things is between those who have it in common, as a table is located between those who sit around it; the world, like every in-between, relates and separates men at the same time. The public realm, as the common world, gathers us together.[3]

"In between" us are interests (*inter-esse*): those things we share, but about which we dispute. The things between us can be quite different: not a table, but routing tables separate and connect us on the internet. The network connects and separates us, and it does so with a speed, scale, and calculative automaticity that has been built both to achieve what is common to us, and to attempt to appropriate it.

Is speaking-writing on the internet a form of speech and action? Is making something (code, software, models, licenses) on the internet a form of speech and action? If so, what does it mean that either can be done by "finding the right words at the right moment"; and for that to be the case, how could it be something anyone could participate in? This was, for a brief moment, a hope held out by the internet we could have had. In this brief moment, a lot of things were begun, a lot of people found themselves in the right moment, with the right words; and regardless of the content of their words, what they initiated was action. The result, however, was the creation of the condition in which we find ourselves today – and it is not, for the most part, a good condition.

Speech and action on the internet we could have had took many forms, but was also accompanied by a lot of *fabrication.* Arendt is at pains to distinguish *work* (and labor) from *action.* Because it is a technology, the first impulse is to say that the internet is nothing more than a work: something made, durable in the world – what Arendt described as the stuff that makes up the human world (as opposed to that which is given by the earth).

If the internet enabled both a technical thing and the formation of new relations, then it is not clear whether the infrastructure itself or the things "on it" are works or actions. Are they more like books and plays, music and sculpture, or are they more like interactions, ways of relating, forms of domination, and opportunities for rebellion? Weirdly, they are both. The lively, animated nature of computation is different in kind from the work of humans in dialogue with each other, but it also animates that dialogue and interaction in new ways. This is what makes it hard to grasp as a political object.

It is possible that Arendt's approach will not help us understand what happened to the internet we could have had. It goes without saying that the internet we do have is not, or no longer is, the space of appearance. There is nothing approximating that kind of politics – democratic or popular power that Arendt theorizes – in the world of Big Tech today. There is strength, and there are various forms of violence, but power, as she described it, seems to have disappeared.

Nonetheless, the theory of the case I am presenting here rests on this idea: that the internet we could have had was a space of appearance in which all manner of people acted, created things – new beginnings, processes, that unfolded in ways that were unpredictable and uncertain. We tried to solve problems, and we made things worse. Now we live with it.

According to Arendt the only thing that would bring such processes to an end is forgiveness: "The possible redemption from the predicament of irreversibility – of being unable to undo what one has done though one did not, and could not, have known what he was doing – is the faculty of forgiving. The remedy for unpredictability, for the chaotic uncertainty of the future, is contained in the faculty to make and keep promises."[4] If there is anything to be redeemed in the internet we now have, it will take better understanding of what the promises were, and whether the debts they created can be undone.

§

Thirty years after *The Human Condition*, Jean-François Lyotard wrote a technical report to the Quebecois government that became most famous for its title: *The Postmodern*

Condition. He called the book a "hypothesis": that the status of knowledge would change under conditions of computerization. It's a peculiar book – knowing nothing of the internet, Lyotard nonetheless saw some things with clear eyes and set them out in lapidary form. He described the fall of "grand narratives" of modernity. He suggested that this was a problem of the legitimacy of knowledge, and at the same time the problem of the legitimacy of politics. Politics, for Lyotard, meant knowing what the rules are, consenting to them, and playing hard. He described his theory in terms of games: language games, game theory, paralogy. Lyotard would have seen algorithms as part of this structure (technologies, the software, the protocols) and that they are also the site of politics. The book ends famously thus:

> We are finally in a position to understand how the computerization of society affects this problematic. It could become the "dream" instrument for controlling and regulating the market system, extended to include knowledge itself and governed exclusively by the performativity principle. In that case, it would inevitably involve the use of terror. But it could also aid groups discussing metaprescriptives by supplying them with the information they usually lack for making knowledgeable decisions. The line to follow for computerization to take the second of these two paths is, in principle, quite simple: *give the public free access to the memory and data banks*. Language games would then be games of perfect information at any given moment. But they would also be nonzero-sum games, and by virtue of that fact discussion would never risk fixating in a position of minimax equilibrium because it had exhausted its stakes. For the stakes would be knowledge (or information, if you will), and the reserve of knowledge – language's reserve of possible utterances – is inexhaustible. This sketches the outline of a politics that would respect both the desire for justice and the desire for the unknown.[5]

Like *The Human Condition* before it, *The Postmodern Condition* is both a philosophical text and one diagnosing a technical society coming into being. It may not be the best diagnosis of our contemporary conundrum, but it is nonetheless a clear document of the moment just before the internet we could have had. It recognizes the centrality of the "market system" as part of governance with and through knowledge; it theorizes democracy and popular power as speech, and as a game requiring knowledge; and it counsels openness, a kind of hacking and leaking (open up the memory and data banks) in the service of democracy (give the public free access). It collapses knowledge and information (but for a reason related to this transition) and it holds out a hope: that knowledge is inexhaustible, and this fact serves the possibility of justice.

In the four sections that follow, all of these elements recur. First the core problem of openness troubles the internet we could have had from the beginning; this is the story of free software and the proliferation of "open everything" (open access, open data, open government, etc.). The conundrum of open source today is that it facilitates a form of "terror" – arbitrary domination – that it was intended to prevent. Rather than ensuring inexhaustible utterances, it has been absorbed into a system for controlling them: not what can be said, but how and through what means it can be said, from the basic infrastructure of the internet we could have all the way to the algorithms governing speech.

Second, the form of arbitrary domination we see today was paradoxically consolidated by the promises of a decentralized network, one that was supposed to be capable of spontaneous order and infinitely open to the future (knowledge is inexhaustible) but which led instead to a cloud of capitalisms.

And third, participation created a new challenge: if speech could be liberated, what exactly would it become? Would it be online, and among an unpredictable and changing collective of hearers, where the memory and data banks are open, and indeed where there are no bankers, no gatekeepers, anymore? Such a liberation would change the nature of the games that Lyotard espied 40 years ago. With the help of openness and decentralization, participation forged a new kind of political subjectivity. Has this changed the very stakes of knowledge and justice as well?

Lyotard was always interested in political subjectivity of this sort. He described individuals as living in a complex series of nodes through which messages of all sorts pass and are transformed, constituting the very social bond.[6] He could not have predicted just how many messages pass through just how many nodes today, or how our social bond is now made up of a nearly infinite series of collectives. He characterized the social bond, this "knowledge diversity," as a language game. "To speak is to fight," he said – a generally agonistic but nonetheless joyful process of statement and counter-statement that engaged people as endlessly capable of making new statements. They would make unexpected moves in a game – not just repeat polarized arguments – moves that would be predicted by even the most average player.[7]

Without that "inexhaustible" reserve of knowledge, he said, society would simply be governed as a giant bureaucracy obsessed with perfecting its input/output ratio, its "performativity." This he did not want: "the nineteenth and twentieth centuries have given us as much terror as we can take," he wrote; our goal should be to resist this terror wherever possible, to play the game well.[8]

Lyotard distinguished three games: true/false (denotative), just/unjust (prescriptive), and efficient/inefficient (technical). To speak is to fight . . . in one of these games. To argue according to truth and falsity is one kind of game, governed either by logic or by some version of scientific demonstration and evidence. To argue according to justice or injustice is to appeal to ethical intuitions, to an indication of harm, or to principles and faiths. Most interestingly, to argue according to efficiency

– the technical game – is to show a better input/output ratio: more power for less money. All these games overlap, but all of them depend on knowing both what the rules are and who controls the information.

To these three games, we might add others today. The "visibility" game is one that will emerge in the sections below: faint/visible is a game we play with the algorithm. Will my speech be heard? Will it go viral? How much must I change what I say to be heard? A "scalability" game is also in evidence, similar to visibility; it is a variant of Lyotard's technical game but instead of efficiency/inefficiency, it is played according to a logic of scaling/failing – a very Silicon Valley game to be sure – but one anyone can play as soon as the rules concern the size of an audience, the growth of a reputation, or the reach of "impact." These are the games of speech and action we play today.

Lyotard used the word *terror* to refer to a very specific version of the technical game: "Whenever efficiency (that is, obtaining the desired effect) is derived from a 'Say or do this, or else you'll never speak again,' then we are in the realm of terror, and the social bond is destroyed."[9] Lyotard is aiming at something like what philosophers of freedom call "arbitrary domination," which is an important concept in the context of technological development because it concerns the way control over a technology arbitrarily limits what others can do with it.[10] Technologies that come with the promise of freedom, convenience, efficiency, or even joy, but which provide those things so long as one does not use another technology, or change the technology in any way, or fail to continue to use the technology even if one does not want to – these are the things that create "lock-in"; it is addiction and engagement but also entrapment, captivation.[11] "Terror" describes the position of the subject, not the action of the entity in control; it describes a closed system, but one whose boundaries are never entirely clear, hence the terror invoked in trying to stay within them, and the subsequent lack of freedom.

The opposite of terror is openness then; in particular, it is the "open system" that is capable of generating new knowledge or new rules. In this respect, openness means something more than availability or visibility, it means generativity or extensibility, precisely the elements associated with the internet we could have had. This is, therefore, a good place to start.

If Openness . . .

Once upon a time there was a distinction between writing and making. Software made trouble with that distinction. It didn't start with software, writing always had a "performative" side; and mathematics was always "unreasonably effective." But software, as it emerged in the late twentieth century, quickly grew into something in between craft and engineering, something between the art of writing and the handwork of sculpture. Like writing, software is meaningless without readers, but this too was blurred. On the one hand, there were "users" of software, and on the other, a community of people who made it, thought through it, could read and respond to it – to the source code or to the data structure or to the function and system calls. It was out of the latter that free software emerged.

There is no internet we could have had without free software. The farther away we get from this movement, and the less well understood it becomes, the farther away we are from the internet we could have had.

Free software was a utopia. There is no disputing this fact, though it all hangs on your definition of utopia.[1] It was a "concrete" utopia in which real participants imagined a set of relations that could obtain between people – primarily concerning property – and then set about trying to make that world a reality.

Free software is not – or more precisely, is *no longer* – open source. Once they were just two names for the same thing. The gray chatter about this, to say nothing of the scholarly literature, stumbled on the fact that free software and open source both used the same licenses. I myself made this argument about them because one of the things scholarly debates in the 1990s and 2000s were telling us was to pay attention to the technical and material details, not only the linguistic, ideological, or theoretical claims. If you look to the details, it appeared as though free software and open source were identical technically, but differed only in terms of the words one chose to use in order to refer to it, whatever it was.[2] It turns out naming makes all the difference.

The stakes of "Free" vs. those of "Open" are different, but they shared a commitment: the idea that there should be no single right way to do something. There is no one set of values or ideologies or standards that should dominate: plurality is good. A permanent openness to any possible set of values, and a mechanism for contesting them. In this respect, it participated in what I describe in the section "If Decentralization . . ." as the trouble with decentralization and emergence: a faith in the power of spontaneous order, rather than in the centralization and top-down control of anything.

Free software and open source shared, for a while, a core commitment to a "space of appearance." It was simultaneously an arcane space made of technical artifacts like compilers and operating systems and software version control systems, and also the speech-writing and relation-making of making these things together on the internet. Because to maintain this space was to preserve the idea that the technical is contestable, just as in a democracy, ideas and proposals and laws and judgments remain ever contestable. An incontestable technology is a form of arbitrary domination. At the outset, both free and open source software shared this idea. They were backstopped by novel documents and processes: ideas like the "Request for Comments" or the Internet Engineering Task Force and its procedures for testing and implementing standards.

Free software was different from open source in its ideology and in the theory behind it, but it was also different aesthetically and politically, facts which had a grounding in material differences involving what it meant to create a bit of software in the first place.

The GNU General Public License (GPL) is justifiably the central achievement of free software, but it does not define free software's ethos. This famous software license, created by Richard Stallman and Eben Moglen and celebrated for its cleverness, is one of the greatest and most curious (and now invisible) artifacts of the information age. Its influence is undisputed, having spawned a bestiary of licenses that build on and differ from it in myriad ways.[3] But the reason why free software existed pre-dates and is larger than the GPL. Free software's *raison d'être* was to create high-quality software in a communal environment of camaraderie, sharing, and debate. The GPL was designed to protect that community and its works from property enclosure and secrecy. The aesthetics and politics of this desire preceded any particular thing the software would do, and were taken to be independent of the uses to which software could be put. Just as a house can come to shelter a saint or a murderer, software was not in itself a good or a bad thing. Free software advocates believed that *how software was made* and *how well it was made* could turn murderers into saints or vice versa.

The utopia of free software therefore had a clear moral vision at its heart: that the ability to make software was a kind of core, human activity that could not be alienated (in particular, it was deeply offensive that it could be owned at all) and that making beautiful software would make the world a better place. The free software movement was essentially the Arts and Crafts movement under new management.[4]

If there is a spiritual guide to free software, it was not Karl Marx and not John Stuart Mill, but William Morris. Morris' socialism was grounded in craft: aesthetics very much preceded and determined politics or theory for him. For Morris, a well-crafted textile, a decently printed book, a nicely carved piece of furniture was not an excess but a necessity. Modern capitalism had inverted the possibility of a utopia of beautiful objects for everyone by making everyday objects cheaper and duller. Moreover, capitalism elevated that which could be made beautiful for everyone into luxury goods outside the reach of anyone.

In the world of free software, the heir to Morris is not Stallman but Donald Knuth. Knuth's religious devotion to the craft of software is legendary, exemplified both in the mammoth multi-volume tome *The Art of Computer Programming* and in the famous typesetting software language and toolkits TeX and LaTeX. Among other things, Morris and Knuth have very strong, and curiously opposite, aesthetic commitments to the specific but very arcane issue of letter-spacing in typeset pages.[5] But they share a faith that something so minuscule might be scaled up into a politics of making and manufacturing, and a belief that for humans to thrive, aesthetics should precede politics, and therefore manufacturing, and not be subordinated to it. Ironically, the origin story of free software, concerning Stallman's frustration with proprietary software, was that he wanted to improve a driver for a printer (perhaps to adjust letter-spacing?) but could not access the source code to do so. Writing as craft binds these experiences together, and all three men (and yes, it was a utopia of men) shared a sense that craft and community were connected.[6]

Regardless of whether such spiritual resonance is acknowledged or not, the core writings around free software – and open source – tend to emphasize the ability to experiment with code, to improve it, to share it with others, and most importantly and magically, to see the benefits of well-crafted code distributed equally to everyone. There is never a concern with the idea that a poor improvement or broken code would be distributed equally to everyone; this would be, according to the ideals, fixable by whosoever could modify, improve, and redistribute it. By contrast, large organizations, relying on the crutch of intellectual property law, were enabled to create cheap, poor-quality software which they could force their consumers to use, which they could fail to maintain, and, ultimately, for which they could charge a never-ending rent.

When disputes in free software would arise as to what "best" or "most elegant" means, it would be precisely the presence of a community online, willing to argue and debate, that would sort out the right course. In the few cases where irreconcilable differences of opinions stultified the beautiful, the very nature of software was to provide alternative possibilities: to extend plurality rather than reduce it. For most people engaged, the presence of more than one great text editor, for instance, was not a problem to be fixed but an opportunity to let at least two flowers bloom, and a playful sense of competition to thrive.

The idea that there could be more than one best way to do something was consistent with the "self-leveling playing field" (cf. "If Decentralization . . .") and at least some versions of free market thinking. To be sure, it was also socialism in the core sense of preserving the dignity of work, but combined these two supposedly opposed ideas in a resistance to the idea of a zero-sum game, or a winner-take-all, network-effects model of creation. The very idea of "forking" was central to the technical design of licenses: one could always, on certain conditions, take the source code of a free software project and make something else with it. The resulting improvement could then join the world of alternatives, competing, collaborating, evolving, improving.

Such was the ideal. It was also a real puzzle at the time. Mainstream economists were puzzled above all: "How can you make money selling free software?" As noble as the free software ideals were, no one would oppose them to the idea – the necessity – of making money. During this period, many different kinds of people actively contributed to the creation of free software that became very, very widely used on the internet. Many were university employees, ranging from professors to staff and students. Some were in corporations or research labs like Bell Labs. Most saw what they did as publicly funded research that belonged to the public, and therefore should be made available for any purpose (including for people who would commercialize such goods!). Many others were independent professionals, working on such projects on the side, especially on projects that supported the core infrastructure of the internet. Their passion revolved around making the tools and technologies that could take advantage of the magic of the decentralized, emergent, spontaneously ordered network of networks.

Free software was never directly about making money though. It was about collectively rebuilding something, even if that something was ultimately understood to be supportive of market capitalism. Free software had no theory of an honest wage, it had no union nor a collective nor a cooperative to support the vision of an economy implied by free software. It existed naked within a high-tech economy in which the pressures to exploit good ideas and turn them into products or services was intense. Even though the internet we could have had, up until the early 1990s, was "non-commercial" in a strict sense – government-provided, run by universities – it was already soaking in a bath of capital, intellectual property law, and dreams of a new economy. The people who made free software did so "on the side"; they paradoxically worked day jobs to support their utopia of work.

Debates and discussions in this period turned around this paradox. There was the widespread moral outrage at the idea of enclosing communal property to sell to those who had contributed to it. If software was in "the public domain," it was there for everyone to use. Selling it to an unsuspecting someone was a morally corrupt thing to do – it violated a cultural norm. Outrage was balanced by outrage where the question of just compensation for creating software was concerned: the professoriate in the universities and the weekend tinkerers whose day jobs were in the tech firms both failed to appreciate the emergence of a new class of people who started to see themselves as software professionals (programmers, developers, architects, designers – the terms proliferated through this period). Such people saw not only justice, and dignity, in the idea of being paid a wage to create software, but also the equally important sense of opportunity to create new businesses and new economic growth out of what they had built. Everyone from their sixth-grade teachers to Al Gore was telling them to do so.

In the mid-1990s there appeared an intense commercial interest in, and mandate to make money on, the new internet infrastructure. Some of the people making free software saw their freely given work being exploited for money. On the one hand, some people demanded the right to make money on the software they created, while others (sometimes the same people!) demanded that no one else make money off the software they had created. People watched as businesses used software they had created in order to make more money (and get more attention) than the makers ever had in the first place. Most of this software, to be clear, was openly available and visible to all, because of the innovation of the free software license.[7] One could read the rise of free software as an attempt at self-protection by a quasi-academic community, just at the time when universities had begun to aggressively insist on the commercialization of the work that they and their funders paid them to create.

This convoluted mess found precise expression in the differences in licenses. The GPL and it's famous "copyleft" clause (negatively referred to as a "viral" license, before virality was a good thing) explicitly forbade anyone from using the software to make software unless that subsequent software was also made freely available. In contrast, a range of licenses dropped this clause in favor of the idea that "freedom" included the right to make choices about how to do business: rather than releasing software under the same license, it allowed people to charge money for it or release their changed software under a different, more restrictive, license.

Although arcane, the debates in this period about software and licenses were essential to the internet we could have had. They gave shape to a new understanding of property. They resulted in a range of expectations and practices around how to talk about, use, and enforce free software-style licenses; they included a range of new licenses that would be applied beyond software – most dramatically the Creative Commons licenses that covered writing, music, video, and other kinds of digital content that could be remixed or re-distributed. The Creative Commons licenses mimicked and extended this division by offering a choice of license terms that limited users' ability to (or gave them different freedoms to) do certain things (a "non-commercial" restriction for instance, or a "share-alike" restriction modeled on copyleft).

"Free" software always comes with the adjunct insistence that it is "free as in speech, not as in beer." There was never any ambiguity that this movement was about a particular definition of freedom, steeped in that model of human making and dignity – the freedom of *Homo faber*. It was not necessarily a freedom associated with equality and justice, but with speech, action, and making. The freedom in free software was not negative, but positive and republican, after a fashion.[8] Among software programmers and pundits in the tech world, a range of rather idiotic debates were mounted about whether free software was "business-friendly" or whether the word "free" was inherently confusing. Non-English speakers tried out "Libre" software, since the collapsed distinction was not a problem in other languages. The term "open source" emerged as the alternative in 1998, coined by Christine Peterson, from deep within a key site of internet futures at the time, the Foresight Institute.

Open source was "business friendly" because it was about creating a "new economy," one in which growth and productivity were not tied to per-unit sales of software but to the (always deferred) promise of quality, and the power of having an impact. Open source offered the benefits of not having to reinvent the wheel; concern with freedom or privacy was easy to ignore because "openness" was about stability, quality, and the ease and rapidity of modifying software.

The freedom of the consumer (choice) was more important than the freedom of the craftsman (work). Consumers included entrepreneurs, start-up companies, and, ultimately, large corporations themselves. By the time of Snowden's revelations, the very largest corporations were already the most enthusiastic stewards of open source.

However, the rise of open source was not simply a capture by capitalism. By its nature, it was open to contribution and remaking; many different kinds of people – academics, educators, musicians, writers – embraced open source and its non-software variants because they traded in a different currency: impact. The values associated with open source, beyond the narrow questions of property or transaction, came to be associated with the moral authority of speaking in the public sphere, influencing debate, or having that increasingly elusive impact.

As the term "open source" proliferated, the ethos of free software changed into that of open source: from the virality and protection of the communal production of software whose main goal was self-actualization and community cooperation to the generalized sense of "openness" as visible source code, generating innovation because it was available for building upon. From a commons in the sense of Ostrom, back to one in the sense of Garret Hardin: no longer a small-scale system of cooperative management but a large-scale system of automated incorporation and expansion.

Free software persisted and differentiated itself. It became more strongly associated with non-commercial activities, more strongly devoted to issues of privacy or encryption, more focused in many cases on specifically left or progressive causes – because it was about freedom first. Terms like "FOSS" and "FLOSS" emerged as ecumenical labels in academia and activist circles, but in any commercial environment, the term "open source" dominated by the 2010s. Slowly, a kind of mitosis occurred: free software and open source became quite different entities; as the internet we could have had morphed into the internet we do have, free software retreated as open source dominated the landscape.

Open source became something else: a large-scale infrastructure of source code sharing and hosting; first in relatively distributed ways (this is the story of "git," the software necessary for version-managed source code), but then suddenly much more centralized in repositories like GitHub, now owned by Microsoft.[9] As in the case of Amazon Web Services (see "If Decentralization . . ."), it provided the convenience of a centralized, zero-capital investment service, where rent extraction substituted for any other kind of transaction.

License choices, software release cycles, frameworks for managing open source software, conferences, professional societies – open source became the de facto development practice in the high-tech industry throughout the 2010s. Every major Big Tech firm now open sources its software: Apple, Google, Facebook, Amazon, Microsoft. But the irony is that this technique for creating an open commons in the internet we could have had has come to produce the exact opposite: the most powerful monopoly corporations since the Gilded Age. What happened?

. . . Then Google

Many of the people who advocated for open source firmly believed that it would change the way capitalism worked. Most of them believed it would make capitalism better; those who clung to the original ethos of free software continued to claim for it a form of cooperativism or anarchism, which still exists in places.[10] The virtues of open source were said to exist independently of the power or cravenness of any particular corporation or government. Making source open (or making data, hardware, or anything else open) brought the power of sunshine – distributed and robust. Software would be better, innovation would move faster, and more people would be enabled to participate in the creation of wealth.

Early on, free software advocates like Eric Raymond and Bruce Perens worried that the "ideologies" of freedom and dignity in software creation were driving corporations away from embracing open source. Struggling start-ups pioneering free software business models (like the Red Hat or SuSe operating systems) proposed new kinds of economies, built differently than in the past. In 1998 Netscape open sourced its browser code in order to compete with Microsoft; Apple did the same with its Darwin operating system (a variant of UNIX birthed during Apple's dark years without Steve Jobs) two years later. The former did so as a last ditch move to gain market share without monopoly power; the latter did so because they maintained strict control over the hardware that Darwin ran on.

Nonetheless, most of the largest corporations were slow to embrace open source for a more basic reason: it wasn't clear yet how to control it. It would take a minute for corporations to realize how to do so. The slow realization dawned throughout the 2000s: if corporations could contribute to open source through their own employees (whom they could manage), then it would be possible to influence, if not control, the design and development of software, while at the same time gaining the good will of all those people who asserted that open source was good for the industry and consumers, not just owners and management. Daneese Cooper, one of the few women in positions of influence in the open source world, was one of the first "open source evangelists" (first at Sun, then at Intel) to work with corporations to embrace the model (starting in 2000). Other formal positions devoted to transforming corporate practice to embrace open source emerged in this period.

Google is a good place to observe what happened as a result, and, in particular, in the way it managed the browser it created: Chrome. In an article on this topic, Bolun Zhang and Davide Carpano give a detailed account of the process by which Google was able to convert open source software into what they call "logistical power."[11] Their account is important because it differs from other critiques of free and open source software, and is more empirically grounded.

As Zhang and Carpano explain, academic scholarship on free and open source software has generally been confused about what it is or does.[12] For most of the past 20 years, critiques of free software or open source by academics have focused on trying to explain two aspects of it. The first is that open source relies on "free labor" and that this is the source of the surplus value it creates.[13] The second is the idea that it attracts users by providing a free product, and then traps them and extracts data from them. The first claim is true in part and, by the time of the "techlash," a small core of open source developers around the world had started to make new sense of their own exploitation.[14] But most open source developers were, however, either academics or employees of corporations, not exploited members of a working class. The second claim mistakes the work of creating software for the uses of it by others (indeed, it mistakes the "utopia of work" beloved of the original free software movement for the "gig economy" that resulted from the rise of platforms; the former is about the work of making software, the latter is about using software to organize the work of others).[15]

Neither explanation, however, is relevant to the power that open source provided to corporate strategy. Open source did not provide free labor, or extract valuable data, it enabled the control of the standards by which anything else might be done; in particular, it enabled control of the standards-setting process by flooding projects with their own designs, labor, and participation. The result is a product that Google owns on the inside but not on the outside, as it were: an open source browser engine that can be freely and openly used by anyone, based on a set of standards that they ultimately control. By combining the openness and availability of software with the ability to force the standard, a new kind of power emerges.

The basic structure of corporate open source strategy involves several steps: first, agreeing to participate in cross-firm open and collaborative production of software, whether that is browsers or AI software, or programming languages or any of the less glamorous pipes of the internet necessary for its function. Second, it involves hiring and providing the labor necessary to create the components and processes that a corporation wants to see in that software. And third, the twist of the knife: using their financial and political power to either subtly or overtly control the direction of development of an *open* standard, a protocol, or a technical object like a software programming language or framework, to match their own strategic goals. The significant advantage here is not just that a single corporation controls a technology – they now control an *open technology* that all their competitors will most likely have to use as well.

One example of this process is that of the Chrome browser (and its core engine, Chromium). Browsers almost inevitably became the de facto operating system of the distributed internet. Early on, it was clear that whatever the browser might be, it would be the interface for a great deal of computation, storage, interactivity, and communication. Open source has, in fact, enabled exactly what was promised: a range of different browsers, many of them forked from previous projects. A cornucopia of choice exists for the consumer, especially after the conclusion of the Microsoft anti-trust case and the US government's demand that the "browser market" be one of choice.

By about 2015, all of that choice consolidated into three huge "browser engine" ecosystems: Gecko, maintained by Mozilla Firefox; Webkit, maintained by Apple; and Chromium, maintained by Google. In all three cases, "maintained" is precise: they did not create these ecosystems, but only contributed to them as open source projects – contributed rather than built, shepherded with a rather draconian hand perhaps.

Open source projects – especially those that form the core infrastructural parts of the internet like the Linux operating system, the Apache ecosystem, programming languages, or a case like the Chromium browser – have long been overseen by a variety of organizations, mostly non-profit foundations, occasionally by corporations themselves. Some are "standards organizations," whereas others are foundations or non-profit organizations. The idea, central to the internet we could have had, was that openness and spontaneous order would be better solutions to the problems facing society than a world of closed corporations with proprietary, secret, or otherwise unaccountable software and technologies. Standards support the idea of limiting the power of individual firms. And in a world of relatively weak anti-trust policing, standards became the last great hope of creating those limits. In essence, such a hope places standards and the organizations that manage them (including foundations and non-profits) directly into the role of democratic institutions. These institutions, however, were never designed to be democratically robust, nor were they up to the job of defending democracy through the medium of the technical standard.

Regardless of how these standards organizations and foundations are governed, membership is often dominated by employees of relevant tech firms. This is essential because these are complex technical objects best treated with care by people who know them well. It is also a conflict of interest: employees in these organizations represent their employers and not any neutral or general sense of technical efficiency or excellence or any other kind of values supporting a democratic restriction on the rule of corporations. Whether or not a firm coerces their employees in these contexts, and with what managerial techniques, happens far outside the public eye and even the oversight of the standards organizations or foundations themselves.

Google has long dreamt an old dream of the internet we could have had: the idea that the decentralized internet is a kind of giant computer made of diverse technologies and people. Rather than just a network, it is an architecture of computation. As such, it would need its own operating system and the organization controlling that operating system would hold the keys to the kingdom.

One key to such a dream, as Zhang and Carpano point out, has turned out to be the browser engine and its technologies. Not simply a single piece of software, the browser allows many different kinds of technologies to hook together through Application Programming Interfaces (APIs). Chromium has become "an API production line" through which the capabilities of the browser-as-operating-system are proposed, tested, and then implemented. These changes exist not only for Google, but also as standards that all three browser engines must either implement or fail to offer consumers (indeed, failing to offer these conveniences can be both a source of incompatibilities and, in the case of Mozilla, a broken promise to their consumers). Google's size and density of expertise has allowed them to lead this process, often steam-rolling standards committees in the process: Zhang and Carpano quote a Mozilla engineer saying "Google proposed new API standards so quickly that Mozilla could hardly follow, and this diverted the already limited development resources that Mozilla has." Or in another case, Microsoft was unable to keep up with the changes, which "led Microsoft to drop its own browser engine in 2019, instead opting to build their Edge browser from the Chromium base."[16]

Google's ability to control the standards throughout this open source ecosystem tests the limits of other organizations' commitments to privacy and security (for instance, Zhang and Carpano detail the case of WebUSB, a technology that allows the browser to access a USB key on a personal computer). This process need not be understood as a kind of coercion, or corporate malfeasance. Rather, it is a species of arbitrary domination because even though the open source development system is supposed to allow people to fork projects and develop alternatives, Google's role in the browser engine ecosystem makes such a choice untenable for the other players. It is a counter-case to something like the Linux operating system, which up until now has remained a single, remarkably widely used operating system for the internet. Ironically, that success is due to the iron hand of its original maker, Linus Torvalds, though it is certainly a matter open to debate how much that operating system has remained open, as opposed to following a design path driven through Torvalds, but by the Big Tech firms' interests.

Zhang and Carpano suggest that Google has pioneered a new strategy for controlling the tools and technologies that can be used on the internet; rather than simply owning these technologies, or using their power to approve or disapprove certain uses, they have sought to "reshape the topology of the

Internet in a much more subtle way." Zhang and Carpano end, happily by reference to Arendt, hopeful for more spaces of appearance – in particular, those standards organizations and foundations forming the "agora" where one might find "the right words at the right moment" to assert a form of popular sovereignty. But, they say, "that day will only come after we have devised new forms of Internet governance capable of accounting for concentrations of logistical power."[17]

Google's perverse new authority emerged out of sites of attempted popular sovereignty. Although most people would not consider a standards organization anything like a public forum, a town hall, or a ballot box, they have been made to serve the same democratic purpose and they have ultimately failed to do so.

In addition to forging a new direct, subterranean form of power, one that enables vast large-scale systems to be tuned to the interests and future prospects of the largest tech corporations, the success of open source ultimately camouflaged this work with a dimly lit morality (openness, not freedom; a utopia of convenience, not one of work). Larger and larger firms began to tan in that low-wattage moral glow. They could signal to government and academics that they had the best interests of the economy and society at heart and simultaneously signal to investors that "open sourcing" a product was a step toward capturing the subterranean power of a segment of the tech world at large.

One need not even accuse this strategy of being evil: it simply scaled up the promises everyone was making for openness. Openness became the very keyword of the millennium, for a while, generating a brood of offspring: open standards, open source, open access, open data, open science, open hardware, open content, and so on. All of this openness was offered not out of generosity but because its powers were said to be significant: it enhanced transparency; it ensured scientific review; it allowed anyone to find flaws in an argument or a piece of software; it generated innovation; it facilitated economical reuse; it ensured accountability; it smoothed out markets in information; it even, weirdly, enhanced privacy by making the rules and violation of those rules more visible. It had "impact."

All these values are, however, downstream of the ability to control the technical ecosystems which provide the "open" things themselves. It hardly matters if your article or book or data or hardware are open, if you can only access it at the pleasure of the sovereign and only on his conditions. The irony of openness today will be clear to any contemporary user of the hyper-secret, ultra non-transparent, privacy-invading, and, so far, totally unaccountable OpenAI tool ChatGPT (or any of the other models). It requires neutron star densities of cynicism for OpenAI to retain its name as anything other than a brand. It is a good sign of how far away the internet we do have is from the one we could have had. These companies still trade on the idea of openness: they "open source" their models, which is a monumental fudge.[18] Meanwhile, the models and infrastructure of AI are paving the earth with data centers and a utopia of not-thinking, all "open."

The idea of openness nonetheless stubbornly retains the moral valence of a basically good idea. Its opposites – secrecy, opacity, back-room dealings, and asymmetric information – all carry the taint of manipulation, cravenness, and unaccountable power. No longer does the term "open source" mean anything other than a bid at massive, logistical, and infrastructural control. No one should imagine a public commons, available to everyone, nor should they hear "a system for producing the highest-quality software" or for ensuring security, or for any of the things that it was initially promoted to provide. The same goes today for "open data" or "open hardware." But this is *the world we now live in*, and not the internet we could have had. There is and was a digital commons that resulted from the internet we could have had, but it is no longer available to us on the terms with which it was created. In the future, it will have become nothing more than a giant matrix of vectorized tokens and generative pablum and no one knows yet what that will do to the society, culture, or the economy (maybe we will be saved and it will become the AI we could have had).

In the internet we could have had, free software was a beautiful, latter-day cooperativism, in the spirit of Chartism, or William Morris, or the dreams of Fabian socialism; a utopia of work, a world of craft and quality, aesthetics before money. It was also a pedagogical experiment, not so different from Tagore's experiment at Santiniketan or Paulo Freire's pedagogy of the oppressed: small-scale but everywhere necessary for liberation. Most people will scoff at that claim, but many on the internet we could have had recognized that such ideas needed to find a place in the high-tech worlds of the future. What free software shared with these other experiments was close study, contemplation, and mastery of a craft as routes to liberation and social cohesion. What emerged out of it was the opposite: scalability.

The story of free software and its descendants has always been a story of keeping the commons safe from destruction. The reciprocal (viral) clause of the GPL was narrowly focused on keeping everything in the open commons of a collective who would maintain it. It was not at all focused on scalability; the goal was never to scale up the utopia of work and enforce it at every level, or determine the tools that everyone would use. There were those who embraced the indeterminacy of letting the work of their hands be recombined and scaled up for unknown purposes, but there were also those who saw it as a threat, a need to tread very carefully because it was not clear that openness always leads to the best of all possible worlds.

Free software was also initially designed for individuals and small groups, not corporations or states. It is only by imagining the corporation as a kind of person that such a use can make sense – and then only in a legal sense. But corporations' goals are different than individuals' goals. Individuals are self-reliant (in an Emersonian sense), and their goals are to manage relations with a small number of other people with whom they interact. Corporations are super-human, they are experiments in forms of action and control that no individual human could conceivably engage in. But a corporation cannot really be together with another corporation in a space of appearance; it cannot avail itself of the kind of power that humans can; strength, violence, rule are what they were designed for, not power.

Out of these ideals in free software, open source emerged as an attempt to implement the fantasy of a New Economy, and, by fantasizing it and animating it, made it come true. But it was nothing if not a revenant old economy, the very one we'd hoped the internet we could have had would vanquish, yet it grew. An economy of monopoly power now capable of subordinating a wide range of economic activities (one kind of capitalism) to the logistical and infrastructural power of the Big Tech firms of the internet we do have (a second capitalism). It is this mystery to which we turn in the next section.

If Decentralization . . .

Once upon a time, history came to an end.

Imagine, or recall if you must, the 1990s. All the cool kids (and a bunch of people who wore business suits) are talking about globalization.[1] What would it be? It didn't yet exist (except, said historians and others newly sequestered in university "ethnic studies" departments, it's been going on for 500 years and no good has come of it yet). Globalization is "multinational corporations," it is flows and scapes, it is global cities, it is supply chains and containerization.

The story of the internet we could have had begins among attempts to think with and against globalization and the problem of "free markets." The decentralized, spontaneous, power of the internet we could have had birthed experiments in remaking economic relations: between consumers and producers, among firms, between firms and governments. It was going to create a world of properly free markets, it was going to balance the powers: but that is not what happened.

Globalization was about "information and communication technologies" or "communication networks" before it was about the internet. Globalization was about media conglomerates and their broadcast networks. Globalization was about "bodyshopping" and "offshoring."[2] The internet gave sudden specificity to this vague talk: it seemed an obvious motor for globalization, as a de facto international technology, built on telephones and telegraphy on the one hand, and an international scientific elite on the other. Free market ideology was the flashpoint of the decade – the End of History had happened and the triumph of capitalism and democracy was declared.[3] There was a rush to implement free markets in the former Soviet republics, using freshly applied lessons from the failures of structural adjustment a decade earlier.[4] "Neoliberalism" – which was hitherto a relatively obscure neologism describing certain developments in Latin America – suddenly burst into widespread political use around the implementation of NAFTA and GATT.[5] By the mid-1990s the internet was also a thorn in the side of globalization because it seemed it might allow rapid resistance to flourish: in Chiapas, for instance, under Subcomandante Marcos of the Zapatistas.[6] As a result, "Netwar" was all the rage in the think-tanks of the defense establishment.[7] The 1990s also brought us Zima, a clear, undrinkable soda, and also potato chips that came with a warning about the risk of anal leakage.[8] A lot of things were happening.

"Free markets" were "unquestioned" though; they would be the central and robust mechanism by which both globalization and Olestra could flow. The triumph of the West over the Soviet Union had established the perfection of the free markets under Western democracy. All that was left to do was to sit and wait for the flows to commence.

Say what you might about free market ideology, but there was actually no consensus. There were those who resisted the spread: the internet was one locus of anti-globalization. But there were also those who said: the main problem with free markets was that they actually weren't free enough. Despite the rah-rah of the triumph of the West and the free market, doing business in the West meant doing it among enormous corporations, often corporations undergoing long-term anti-trust investigation and prosecution. These were entrenched powers who held a kind of arbitrary power over all sorts of decisions – and, most importantly, innovation. Corporate seawalls blocked solutions to the issues that still plague us, from pharmaceuticals for emerging diseases to cleaner energy solutions to better healthcare distribution to, especially, global standards and supply chains for expanding markets.

The internet we could have had, therefore, appeared to many to be not just a new technology, a new opportunity – it could replace a bad capitalism with a better one. It could save capitalism from the capitalists.

At the heart of the passion and enthusiasm for the internet we could have had there was the promise of decentralization: distributed, federated, evolutionary, spontaneous, unpredictable, uncertain, chaotic, indeterminate.[9] All this was Good. It promised that the worst excesses of domination, whether by states or corporations, could be resisted and brought to heel by these new technical powers. It was a new kind of limit: like a constitution limits the power of citizens, the internet would be a global limit on unchecked power. Even better, it came with endless possibility as its motor. People could re-imagine free markets and build models of the level playing field that they demanded. But it all came to an end, and that end had a name which was also the name of its beginning: Amazon. Amazon helped bring the promise of endless possibility to an end, by inverting the decentralized internet. A certain dream of the internet we could have had brought Amazon into existence; that same dream explains how they were able to end it. This is a story of both ideology and technology because it is also a story of animating the former with the latter.

The dream of the internet we could have had starts with the resonant contradiction at its heart (see "Reloading . . ."). The internet we could have had would bring into real existence a theoretical claim about decentralized, distributed action, especially – but certainly not only – in economic markets. In a negative sense, it was a claim that capitalism cannot be effective if it is controlled by a single entity; no nation or corporation can or should have access to all the levers. In a positive sense, it was a claim that something unprecedented and unpredictable would emerge from the complicated doings of individuals and collectives who are loosely coupled, flexibly coordinated, and pragmatically attuned.[10]

Academically, the claim rested in a particular style or mode of science.[11] Complexity, evolutionary theory, chaos, non-linear dynamics, emergence: science's bell-bottoms or perhaps leg warmers or maybe flannel shirts, depending on your decade.[12] Distributed, decentralized, dynamic, chaotic, fractal. These academic ideas could have displaced ideas of control, of logic, and of organization. These ideas were successors to mid-century cybernetics: more than just a feedback loop, it was an open system, second order, emergent, out of control.[13] The US National Academies report *The Unpredictable Certainty* captured a version of this claim: "The Internet also demonstrates the remarkable potential (although perhaps the outer limits) for evolutionary standards development and implementation in concert with rapid technological change."[14]

Politically, these scientific ideas resonated with one of the most central stories of the mid twentieth century: an anxiety about the power of centralized control in the hands of a few. With World War II fresh in the memory, and Stalinist Russia only steps behind it, arguments against central planning, and in favor of freedom, free markets, and limited state power, blossomed throughout both academic and political thought in the West.[15] For many scholars, politicians, and intellectuals across the spectrum of left and right, the challenge of constraining centralized power served as a shared goal – whether the solution would take the form of a fanaticism for free markets or one for participatory democracy and autonomous mutual aid; whether the enemy was a foreign nation, colonizer, or one's own over-reaching government.

The internet we could have had – especially its contradictions – *resonated* across the political spectrum because it was as-yet-unknown, untested: it was an experimental system for an emerging style of thought.

The internet we could have had was also *inevitable*. Everyone said so, as they do now of AI; everyone said that there was no choice, that we would have to learn to live with it. But everyone also said, as they do now of AI, different things about what it would be. Whatever it would be, they said it would solve the problems of the present. It could be the alternative to communism and centralized planning, but it could also be the alternative to Empire, and maybe even capitalism itself. Activists concerned about the rhetoric of a triumphant capitalism in the 1990s worked alongside entrepreneurs eager to find value in a publicly provided playground.[16] For the latter, the internet could have been the alternative to excessively centralized and intransigent control within capitalism itself. They saw revolution in the internet we could have had, just not always the same sort as others, and sometimes not with the same predictable political allegiances.

Critiques of corporate capitalism by entrepreneurs under 1990s globalization tended to focus on the intransigence, market power, and deceitful behavior of the largest corporations. AT&T was the very emblem of the monopoly capitalism whose end was foretold even before that of History. IBM had gobbled up the small players in the computer business, and fought an anti-trust battle so epic it rivaled the Standard Oil fight. Microsoft emerged in this era as the new enemy, figured by none other than Apple Computer as equivalent to a totalitarian autocratic corporation dealing misinformation to worker drones ("1984," dir. Ridley Scott).[17]

Engineers, business-people, and academics in the 1990s diagnosed problems of entrenched and intransigent power among large organizations: it looked too much like communism to be capitalism. Technically, the diagnosis concerned any case where large technology firms manipulated software, protocols, or standards to their own benefit, or in order to control the infrastructure that lay just beneath the markets they sought to control. Large organizations sloshed around with monopoly power over large swaths of everyday life; their approach to providing products and services was primarily organized around the technical control of markets based in their own ideologies of expertise and benevolence resulting, happily, in the steady accumulation of wealth to shareholders. The control that large corporations exerted over markets, as well as simple organizational governance issues, looked too much like a version of central planning: increasingly larger firms in control not only of products, but of social relations, and even the future. Accumulation was not the problem, no self-respecting economist would ever argue against growth and profit in the 1990s; openly arguing for socialism in the America of globalization (or the globalization of America) was a very fringe position. The problem, rather, was a kind of centralized decision-making over whole markets, which amounted to a silencing of new ideas, new products, and better solutions; imagine how much more froth there could be in the warm innovative bath that the Triumph of Democracy had just drawn in the global waters of a new era!

In political terms, globalization on the Left was balanced with Reagan-era small government pushes on the Right. The paradoxical demand to constrict the power and reach of the government, while also encouraging the growth of larger and larger corporations, never really triggered the cognitive dissonance it should have in academic accounts at the time. But the pathology of one was indistinguishable from the pathology of the other, and the internet we could have had emerged as a solution.

In the tech lingo of the time, a certain kind of technological ossification yielded great power: lock-in, path dependency, network winner-take-all effects.[18] With these powers, corporations could slow down or speed up not just competitors, but consumers as well. It was a kind of arbitrary power associated with Microsoft and IBM in the industry, and with other corporations in other rapidly computerizing domains like healthcare, scholarly publishing, or banking. Perhaps (said people on the internet we could have had) a commitment to smaller firms and more of them, on the one hand, or participatory democracy and polycentricity on the other, could activate the energy to create, and faith to believe in, an alternative to these twentieth-century Gargantuas.[19]

The battlefields of the internet we could have had spread out industry by industry, each with their own twentieth-century entrenchments. For some industries the internet was irrelevant: oil, gas, mining, could largely ignore it except in the back office and marketing. But for domains like healthcare, education, finance, entertainment, seemingly unquestioned corporate interests suddenly seemed vulnerable to attack. In healthcare in America, for instance, a long-term experiment in market-based healthcare was already under way amid high-profile political debates about single-payer healthcare. "Computerization" of healthcare was a long, tortuous road, made more tortuous by forms of monopoly lock-in. Medical records technologies like MUMPS and radiology imaging standards like DICOM were being centrally controlled and planned by the most powerful hospital systems, like Massachusetts General Hospital or the Mayo Clinic, and the largest corporations like Medtronic, Siemens, or Epic.[20] While the surface debate was all about the privatization of healthcare and the failure of Clinton's single-payer healthcare plans, those swimming the deeper currents recognized that a market-based healthcare system could take more than one form. To the entrepreneurs and engineers and doctors figuring out the internet in healthcare, the key difference concerned whether it would be a marketplace of diverse solutions and constant innovation, or a walled garden of rent extraction controlled by the largest, slowest-moving corporations.

The tactics at stake were technical: games of efficiency and scalability. There were the standards and protocols that made high-tech solutions both possible and extendable; there were intellectual property laws that hemmed in what people could do and how; and there were high capital costs of entry, costs required in order to play by the standards that dominant corporations could set themselves. The internet we could have had was a challenge to these big entities because it represented a form of decentralized problem-solving that could be neither explicitly coordinated nor captured by a single firm. It enabled small innovators to propose solutions and it allowed information and knowledge to flow to the places it was most needed, not just upwards to the top and inwards to the back-rooms. The internet we could have had was not centralized planning, it was emergent complexity.

Ideas of emergent and spontaneous complexity were not confined to the worlds of business and capitalism; if anything, they were coming from elsewhere, from academia, journalism, cultural zeitgeists, and public chatter. Far beyond the business schools and start-up culture of various industries, there were many similar claims about why the internet we could have had was so powerful. Anti-censorship academics and politicians loved the idea that "the internet treats censorship as damage and routes around it."[21] Transparency advocates were fond of the idea that the internet prevented people from hiding information: "taking information off the internet is like trying to take pee out of a pool."[22] Silicon Valley's myth of the garage-dwelling Davids who take down the Goliaths of power was another recension; but so too was the idea that the internet was a liberating tool of community building by those who had been denied access to elite culture, to education, or to the halls of power. Even for those whose memories of the 1960s and 1970s were dimming, the internet was an "appropriate technology," a Whole Earth Catalog, and a participatory democracy, in addition to a solution to the challenges of complex monopoly capitalism.

Even staid observers asserted that the power of the internet lay in its neutrality and standards-based mediation of equality of opportunity. They distinguished the international *de jure* standards approaches of Geneva organizations like ISO from the de facto evolving standards of the internet and the IETF.[23] Where the former reproduced the pathologies of centralized bureaucratic planning, the latter opened up standardization to a flexible, emergent, democratized playing field.

People of all sorts, from engineers to artists, from doctors and lawyers to managers and teachers gravitated to the promise of these ideas: neutrality, fairness, standards, and no doubt a good dose of competition as a motor of virtue. A new model of the economy and its relation to society suddenly seemed possible; new kinds of economic relations; new ways of instituting markets or market mechanisms that would fix old problems in a broken corporate capitalism threatening to go global.

The difference in the new model, however, would simply be that it would make possible for the first time a technically secure form of these values: it would maintain its own openness, level its own fairness and neutrality, enforce its own standards, and lift all competition equally, rather than submitting to the market power or governmental bias of the corporate capitalism of the twentieth century. Truly the rule of law over the rule of men because, now, code is law. It would draw directly on what the internet seemed to offer in terms of its decentralized technical design: protocols that would both ground and ensure these new economic relations. It would also provide a model for the design and promulgation of all such new relations, recursively: a self-leveling playing field. But it would have to be made and animated.

§

The self-leveling playing field came in at least two forms: secular and theological, so to speak. The secular version was "net neutrality" (a termed coined in 2003 by legal scholar Tim Wu), which was a description of the internet we could have had that also went by other names: "end to end" or "stupid networks." The theological version descended from Friedrich Hayek: spontaneous order. It should go without saying that other people had other ideas for economic relations on the internet we could have had. Interestingly, the far right and far left tended to share, if not a revolutionary obsession, at least a disdain for reform: both capitalism and (neo)liberalism were lost causes for many. There were the sea-steaders and other people interested in the total recreation of society – utopians without topos. There were the transhumanists and extropians, imagining either leaving the planet or leaving their bodies in order to perfect new relations; there were also plenty of strands of anarchism, some more serious than others, but most of them interested in reliving an online version of 1970s communalism.[24]

Net neutrality What came to be called net neutrality emerged out of a 1990s debate, among those on the internet we could have had, about the technical inevitability of the design of the internet. The packetswitched, datagram-driven model associated with the TCP/IP protocols and the original design of the internet generated ideas of the "end-to-end" principle, or the "stupid network," each a way of arguing that the value of the internet lay in its decentralization, its ability to allow innovation "at the edges" instead of control over the center.[25] Net neutrality was based on the idea that the power of the internet consisted in its minimalism: no one controlled the structure from above, they could only control the edges of it, and by doing so were confined to the distributed, decentralized production of alternatives that they could but launch into this stupid middle space for consumers and readers and thinkers and others to evaluate and to choose.[26] The internet had no boundaries, it was endlessly extensible, as long as new nodes or networks continued to appear.

Technically speaking, extensibility governed the growth of the early internet: networks of networks could in fact autonomously govern their own activities, yet still join up with and communicate with other networks, as long as they implemented the core protocols correctly. The devil is always in the details, of course: how was it enforced, who determined whether the protocols were correctly implemented? Much of this questioning centered around standards and "open systems": a debate between partisans of a top-down, heavy-handed *de jure* form of standardization (associated primarily in this period with international standards organizations like ISO, often imagined in the 1990s as the bearers of any possible globalization), and a bottom-up, light and loosely coupled, de facto standardization (mirroring the Unix operating system and its "small tools").[27]

More generally, the ideas resonated culturally. There were "folk internet" theories like Metcalf's law of network value, and there were economic theorizations based on ideas of path dependency, network effects, and "small world" networks.[28]

It might be tempting to look back to the past and identify the moment when Big Tech began the process of consolidating their power by controlling the networks through their economic power, but it is historically inaccurate (in the mid-1990s, Google, Amazon, and Facebook did not yet exist, Microsoft was not a platform, Apple was in its doldrums). Quite the opposite was happening: the people who would go on to lead Big Tech were hatching pie-in-the-sky business ventures on the internet we could have had; they did not see a platform, they saw a miraculous system of *endless possibility*. It might have been utopian – a utopia that included a model of increased competition, more free markets, and a limit to monopoly control.

At first, the claims for neutrality were strict technological determinism. The design of a packet-switched network would unleash the power of radical democratic access to the network, with no gatekeepers and pure, uncorrupted, signals of price and market opportunity. Legal scholars and economists, however, recognized that the technology was contingent: it could change, it could be reformatted by existing powerful interests, and it could be made to benefit some rather than others – traffic could be "shaped" on the internet; some entities wanted to ensure "Quality of Service" instead of open end-to-end expansion (especially, it should be noted, those who wanted to use the internet either to ensure critical military functions, or to deliver audio and video in formats that tried to match the quality of radio and television). "Net neutrality" transformed into a normative and legal claim: that the internet *should* be neutral. For anyone selling a theory that technologies are first invented and then society figures out what to do with them, add this fact to the counter case: in net neutrality, something in the technical idea of the internet we could have had was worth preserving *as a political ideal*, not a technical fact. The law would have to insist that the technology be one way and not another; this was a conundrum for a technology whose utopian power was in the endless possibility it presented. Indeed, limiting the internet in this way was a deliberate preference for one kind of business over others: "internet startups" over big media and communications firms.

Over the years, net neutrality became neither a descriptive nor a normative characterization of the internet we do have, but an abandoned legacy. It was just one of many technical facets of the internet that governments could not regulate into existence. The US Federal Communications Commission (FCC) and the Electronic Frontier Foundation (EFF), among others, never stopped advocating for the promises of neutrality, but its practical scope and meaning narrowed and became yet another object, in the US, of executive-branch ping-pong. A few regulators wanted to put the internet into the "public utility" box, and regulate it like water and power, but this never happened. The internet remained, in fact, an open, end-to-end, distributed, decentralized network of networks. But that internet, the one we could have had, gave us something else.

Hayekian spontaneity A thicker, more theoretical version of the self-leveling playing field is grounded in versions of Friedrich Hayek. Though Hayek's name is today a dog-whistle on the Left for borderline conspiracies of eugenics-laced conservative power, his actual writings and ideas held out, for a wide range of people, prescriptions for how the internet we could have had should have been: a theory of spontaneous order, self-organizing processes, and evolutionary markets and institutions. In the 1990s, Hayek's reputation was at its faintest. He had broadened his writing beyond economics into biology and philosophy, and for that reason his ideas were little felt inside mainstream economics, but at the same time resonated with many on the internet we could have had.[29]

Hayek's theories were rich fodder for some of the most vocal engineers in the 1980s and 1990s: Bernardo Huberman, Mark Stefik, K. Eric Drexler, Linus Torvalds, and Eric Raymond made Hayek a warrant for figuring the internet, or decentralized computing generally, as something alive, evolving, unpredictable, and yet strangely safe and desirable because of this.[30] Perhaps the most ardent and mythical Hayekians were Phil Salin and Timothy C. May. Salin invented something called American Information Exchange (AMIX): a bespoke version of the internet as a market, but without the market, or the internet. It did not run on the internet but re-invented many of its technologies; and there was no market in information, but it anticipated many of the ways information would be valued in the future. Salin was both entrepreneur and early tech enthusiast, spinning a mix of technology and folk economic thinking. His ideas and proposals were particularly influential for some strands of libertarian thinking, but also for people like Timothy C. May, the founder of the Crypto-anarchists.[31] May's vision was extreme: a world without governments of any kind, based in radical individualism, enabled by the power of cryptography.

The folks explicitly invoking Hayek were not working Hayekians; none of them were experts debating the legacy of

Hayek; few of them were probably even reading Hayek very carefully.[32] They were not economists debating the socialist calculation controversy in the pages of the *Journal of Austrian Economics*. They were diverse engineers, economists, lawyers, and self-appointed theorists of the internet who saw in Hayek a warrant, or an alibi, or maybe just a legitimating citation for more parochial but nonetheless revolutionary-seeming ideas about an internet we could have had.

Hayek's philosophy was known to most only through *The Road to Serfdom*. But it was both his early work in the 1945 essay "The Use of Knowledge in Society," and his later 1970s work, especially the three-volume *Law, Legislation, and Liberty*, that sounded to many like a blueprint for the internet we could have had.[33] Needless to say, his work was not a serious feature of debates on the academic Left, for whom Hayek was then, and remains today, little more than a convenient emoji representing "neoliberalism," or more recently "cybernetic" capitalism.[34] While debates about globalization and flows raged in the pages of the *London Review of Books*, the techno-Hayek subculture was more likely to be found at the Santa Fe Institute, the MIT AI Lab, or the Foresight Institute.[35] For these folks theorizing the internet as a more-than-technical collective, emerging from the distributed, coordinated action of millions of people, Hayek was a tailwind.

"The Use of Knowledge in Society" appeared in 1945 in *American Economic Review*. At the core of Hayek's insight is that "the 'data' from which the economic calculus starts are never for the whole society 'given' to a single mind which could work out the implications, and can never be so given."[36] Knowledge about prices and values is distributed among all the players in the market, and no central organization could know it all, even imperfectly. It is not even possible to assume that prices stand in for accurate knowledge about some distant part of the system; the problem to be solved is not one of knowledge but of coordination in an economic system. Hayek observed that any sufficiently complicated set of social relations is out of control, in the sense that they operate by piecemeal and local interactions which have a global effect – like evolution in biology – and result in a spontaneous form of order.

The relation between local interactions and the larger scale led Hayek to evolutionary theory; or, as Naomi Beck points out, to an aspirational but untested version of group selection that is "the most contentious and yet the least known part of his intellectual legacy."[37] More Lamarckian than Darwinian, Hayek's appropriation of the idea was an attempt to solve the problem of the emergence of complex, "abstract" society by appealing to the power of spontaneous, emergent order:[38] "He postulated that the rules of social conduct, which underlie the spontaneous order of modern civilization, have spread not because humans understood them or designed them to be effective regulators of collective life, but because they enabled the groups practicing them to expand more successfully and to include outsiders."[39]

The progressive expansion of groups including outsiders rhymed with the idea of a "network of networks": not one big group, nation, race, but a flexible system for aligning different groups. *Law, Legislation, and Liberty* refined these ideas of spontaneous order as an evolutionary phenomenon. That order, according to Hayek, emerges out of institutions and laws which might have been designed for one purpose, but can also produce unforeseeable outcomes – outcomes that are nonetheless better for everyone than attempts to centrally plan them could be. Hayek elevated this economic-evolutionary insight to the level of political philosophy (and, famously, his argument against socialism).

For Hayek, institutions were not designed; more accurately, their design did not determine the outcome of subsequent relations. The resonant contradiction of the internet rhymed with this idea in Hayek. Institutions were the result of the facts that humans are "as much a rule-following animal as a purpose-seeking one,"[40] and that the habits and regularities that resulted from that behavior are embodied in institutions, which, though changeable, are better understood as an archive of evolutionary adaptations than of rational principles for self-governance.

The process he described was also self-grounding: "Mind is as much the product of the social environment in which it has grown up and which it has not made as something that has in turn acted upon and altered these institutions."[41] We are the result of the institutions we have designed, but not of the intentions embodied in them.

With Hayek in mind, the image of the internet we could have had is an unpredictable, self-organizing, and emergent thing, similar to the economies and markets in which humans have spent modernity, but with more power, more possibility. The unpredictability emerges out of a set of design principles, such as those described by David Clark. Hayek's ideas bring something else: that networks of this sort, once we come to live inside them, will change the relations of people – the very "mind" of humans enmeshed in them. The frequent talk of "the noosphere" in the 1990s depended on something like this kind of crypto-cultural evolution.

Hayek recognized that the design of laws and constitutions was a conundrum for evolutionary theory and vice versa. Because his theory was basically teleological, it worked on the assumption that evolutionary forces were always salutary, whereas attempts to institute principles were just minor perturbations in the unfolding drama. He targets, for instance, "social justice": attempts to right a perceived imbalance simply upset the market forces whose genius is, in the first instance, to distribute the most benefits to the most people. But he does not argue strictly in favor of reducing all law to market forces; rather, he simply asserts that such changes do in fact transform society, just not in the way that is foreseen. The ultimate test of this, of course, is the sanctity of constitutions:

> In these circumstances it seems important to ask what those founders of liberal constitutionalism would do today if, pursuing the aims they did, they could command all the experience we have gained in the meantime. There is much we ought to have learned from the history of the last two hundred years that those men with all their wisdom could not have known. To me their aims seem to be as valid as ever. But as their means have proved inadequate, new institutional invention is needed.[42]

Institutional invention was therefore necessary but unpredictable. This is the idea, the resonant contradiction, that saturated certain parts of the internet we could have had. It is not an accident that many of the people involved in the early debates stressed its imitation of constitutional moments (e.g. Barlow's declaration of independence) or were themselves constitutional legal scholars (Lessig chief among them, who declared loudly that code is law, and asked what kind of constitution would be appropriate in this instance).[43] Hayek's work was subtle warrant for believing democracy to be not just the best of all possible systems, but capable of a miraculous form of coordination among diverse elements, beyond the control of any one of them.[44]

Hayek's work promised that out-of-controlness was far more robust and more likely to achieve the goals of economic growth, productivity, or liberal values of equality or freedom, than any single person, group, movement, corporation, or government could promise. No single entity could reliably have all the data, or foresee all the consequences. For Hayek, this new institutional invention should not fall victim to the myth of "constructivist rationalism" which had people believing that institutions were designed by humans and therefore functioned, continuously, according to those designs. Contra this mistake, Hayek urged invention that would recognize the role of human action as an evolutionary force much larger and more powerful than the dreamy commitment to human rationality. But he added that, once invented, such institutions would change human relations: "That erroneous view [constructivist rationalism] is closely connected with the equally false conception of the human mind as an entity standing outside the cosmos of nature and society, rather than being itself the product of the same process of evolution to which the institutions of society are due."[45]

The resonant contradiction at the heart of the internet we could have had is also the contradiction in Hayek's idea: to design institutions that elevate the functioning of a spontaneous order, which in turn yield institutions that are not designed but evolved, in a virtuous circle that includes the very human minds and cultures designing them. Only teleology saved his system; as Beck puts it, "his only strategy to counter accusations of evolutionary fatalism was to claim that growth is inherently good and equals progress."[46]

Unpredictability, emergence, spontaneous order, evolutionary outcomes: the internet we could have had, as a kind of ecology, was a test of these ideas. Packet-switching, algorithmic routing, and the easy extensibility and modifiability of networks, were all that would be needed for the spontaneous order of the internet to emerge. What it would bring, those half-rhyming with Hayek were certain, was an optimality of markets, an increase in equality, and better outcomes for everyone. A tide lifting all ships, with no ship leading; the new *terra nullius* of the noosphere. It was revolutionary not because it replaced an *ancien régime* with an enlightened one, but because it replaced the idea of regime with the idea of a regimeless future, settled and leveled automatically by the power of uncontrolled coordination. Local rules, global order. It was a faith one must have in the power of a system that had been designed, but designed to be uncontrollable. If it sounds theological, then you are feeling the internet we could have had. Hayek's spontaneous order is a theodicy.

Invoking Hayek in this context – especially for ears tuned to stories of vast right-wing conspiracies and cabals of Mt. Pelerin-trekking eugenicists – risks suggesting that the internet we could have had was always already a version of free market extremism, neoliberal fantasizing, and teleological justifications of the survival of the fittest. The re-emergence of these ideas in the blockchain and crypto communities 25 years later, busy composing rules of governance for distributed autonomous organizations (DAOs), is either confirmation of this – or, given the distribution of both radical left and radical right variants in that community – a sign of a deeper failure of these ideas.[47]

It's also the case, perhaps, that one doesn't need the kryptonite of Hayek to explain the strength of the internet we could have had. Elinor Ostrom was also often embraced as a modeler of what could have been.[48] More palatable to an academic Left, but not threatening to pro-market capitalists, Ostrom's analysis of commons was a sophisticated rebuttal of Garret Hardin's famous "tragedy of the commons." Hardin's argument was that, left to its own devices, a central commons will collapse from overuse.[49] That argument buoyed many hard-line free market enthusiasts who argued that strong private property law and unfettered competition was the only solution to government centralization. But Ostrom demonstrated that order in a commons depends on rules and practices that are emergent, polycentric, robust – and incidentally, also human, resonant, and qualitative, not quantitative and programmed – a model that many people thought the internet we could have had would implement.[50]

The competing visions of the internet we could have had – whether gray-suited net neutrality or the tweed of Hayek and Ostrom – were also stories of globalization and its discontents. For a hardcore Hayekian, the internet we could have had would waft down onto the broken, nationalist, ideologically riven systems of post-War first worlds, a diaphanous veil of packet headers and routing algorithms, out of which would emerge an alternative system of globally coordinated action controlled by no single powerful force.

For the net neutralists, it was a more pragmatic preservationism, mixing social justice (free speech and the endless possibility of the internet we could have had) with economic growth for a Washington consensus of international liberal trade and the spread of US soft power and international development.

Not everyone promoting the internet we could have had believed in either Hayek or neutrality, of course. There were those who thought neither went far enough. It was utterly obvious how much power, for instance, the US exerted over this new supposedly global technology. For a system claiming to be neutral, it was obvious on the face of it that it wasn't: Top Level Domains (TLDs) gave the lie to this claim (the US got .com, .org, .edu, while everyone else was defined by their nation: .uk, .fr, .za). There was the obvious geographic and financial centralization of the major venture capital firms in Wall Street and Silicon Valley. Perhaps the most central and egregious concerns of the internet we could have had concerned the global rule of US-based intellectual property law, in treaties like WIPO and TRIPS which forced other nations to "harmonize" with the US rather than supporting elements such as the "moral rights of the author" in Europe. Meanwhile, India engaged in an activist disregard for patents in the name of public health, and China pioneered a "Shanzhai" form of innovation that the US could only understand as criminal piracy in its failure to respect US IP law.[51] Activists argued for no borders one day and anti-globalization the next, looking for a system that would allow the flow of people and ideas without the extraction and concentration of wealth.

For dissenters, following Hayek would mean taking on his "civilization" and his Euro-centrism, which conflicted with activism firmly oriented against it: the post-Vietnam era, the resistance to the neoconservatism of the Reagan–Thatcher era, the Gulf wars, and the misdeeds of structural adjustment in the IMF and World Bank. The internet we could have had could have been an alternative to that civilization, but Hayek might still be correct.[52] Out of control could just as easily mean anarchism: crypto-anarchists, Cypherpunks, and hackers recognized that the "self-leveling playing field" was a domain for exploring aspects of mutual aid, vigorous non-allegiance, and uncovering secrets and lies and technical hypocrisy with the light of the lulz.[53]

In thick and thin versions, the self-leveling playing field of the internet we could have had might have been a perfected anarchy, but the internet we do have is neither. What happened?

. . . Then Amazon

It turns out, *pace* Hayek, that Amazon actually did have all the data in one place, and that was not all.

Amazon was a child of the internet we could have had, a website turned start-up turned corporation turned platform. It was a beneficiary of the "end-to-end" innovation at the edges supported by the internet we could have had. If the bargain of the thick and thin versions of spontaneous order was that, on the internet we could have had, an Amazon would not happen, then how did Amazon figure out how to happen?

Amazon helped transform the internet we could have had into the one we do have when they inverted it and turned the internet into a product of Amazon rather than the reverse. Between its origin as a bookstore in 1994 and the roll-out of its web services and cloud computing by the mid-2000s, Amazon didn't just become the world's largest retailer, it became the market itself; it became logistics and coordination itself; it became the de facto playing field leveler. In the evolutionary terms beloved of Hayek, Amazon was a highly successful invasive species dominating an ecology of fragile but beautiful biodiversity, populated by a range of species with no sufficient defense mechanisms. Or, better, it was both invasive and parasitic, incorporating existing organisms into its vitality and growth. Or, even better, Amazon figured out how to become the ecology itself rather than simply existing within it, ignoring completely the naive models of economic competition among firms in a market that grounded all of modern legal and economic thinking.

The story of what Amazon accomplished is more than one of building the world's largest bookstore or dominating a logistics industry. It is a story of how to change the changing and changeable internet (see the section "Reloading . . ." above). It is a story of cloud computing and of the technologies of virtual machines, hypervisors, and data center engineering, but it is also a story of re-imagining what the internet could be. Amazon created the internet we do have by capturing labor, land, market, and infrastructure all at once, and then figuring out a way to quickly rent it back to the people still invested in the promise of the internet we could have had.

Of course, Amazon is not completely alone in this transformation. Across the globe, there were mimics in Alibaba and Tencent, the rise of the "superapp" and the consolidation of both the real estate of the internet and its integration with cable and cellular industries. And while Google took much of the heat for monopolizing search, Amazon appeared to dominate through diversification.[54]

Amazon started life in 1994 as a bookstore on the internet that, as everyone knows, Al Gore had just invented. When he did so, under the label of a "National Information Infrastructure" or "information superhighway," he was looking at a concrete problem of infrastructural growth. The pragmatics involved in growing the internet ran up against economic limits of concrete sorts. The magic pixie dust of the internet protocols needed a place to land, rails to run on. Investing in the hardware of the network was clear to Gore: it would not do to base the glorious information superhighway of the future on a distributed, heterogeneous network run on National Science Foundation (NSF) peanuts, where Domain Name Servers (DNSs) were a file on Jon Postel's computer, where registering a domain name meant sending Jon a polite email, and where maintenance depended on the part-time attention of graduate students in computer science departments run by academics and long-term research adjuncts like Jon (not to single out Jon, but he was the Hephaestus of the internet we could have had).[55]

Gore's political win changed the internet we could have had in ways that still reverberate through the system, setting the stage for forms of investment in the blue chip-worthy corporations that would make the pixie dust of the dotcom boom work as everyone expected. The work of laying cables, building carrier exchanges, buying up real estate and spectrum rights, and negotiating the costs of moving those trusty packets around, stabilized the core protocols and software of the internet and enabled successes like that of Jeff Bezos.[56]

It's not an accident that the first great platform of Big Tech started life as a bookstore. After all, what was the internet we could have had but a global digital library? It was going to have all the world's information in one place, and if that were going to come true, someone would need to provide the books.

Yet Amazon understood early that it was not a bookstore; it was a place for bookstores to reach larger audiences than their local neighborhood. The internet we could have had was marvelous for just this reason: because it warped time and space in new ways and connected up those otherwise locally rooted communities in a global noosphere, a more-than-market. If only it could be built.

Amazon was smart enough to realize that it was a fool's errand to simply compete on equal terms in a free market of booksellers to reach an international audience of buyers. Amazon's big idea was to make it easier for *other* bookstores to sell through Amazon than to sell directly to their consumers on the self-leveling playing field itself. And why stop at books?

If it were to go beyond books, Amazon needed to become a larger and larger node on the internet we could have had. As it grew, it benefited from the fact that the software and protocols of the internet were openly available; it was one of the earliest and most enthusiastic beneficiaries of free software even before it was called open source software. This was no academic theory – not the legal theory of net neutrality or the Hayekian dream of spontaneous order – it was the trench-level creation of the software that was being made to work on hardware that increasingly came to be seen as a commodity, as interchangeable and replaceable.[57] Computers were wheat, computers were pigs, one chop as good as another. Computers became server farms. If the 1980s was the decade of the personal computer, the 1990s would be the decade of software built out of those farms and fields.

To learn and to harness these dynamics required looking under the hood to see how the internet was configured, and to take advantage of this new knowledge not to create a neutral self-leveling playing field for everyone, but in order to cast spells on the playing field. Amazon did it first.

Amazon not only grew the size of the network nodes it occupied, it started to think about how these nodes were themselves a mirror of the internet. Inside Amazon's data centers, engineers were tasked with finding ways to manage their own network in new ways. They understood the need for "scalability": quickly adding or dropping nodes so that they could meet traffic and usage demands without having to spin up or shut down lots of hardware. Excess space and bandwidth was unevenly distributed, wasted: Amazon's micro-managing CEO Jeff Bezos was gifted at taking advantage of every such shortcoming. Amazon's innovation built on the dreams and plans of the internet we could have had, the freely available technologies being created in order to open up and level out the playing field. Amazon implemented new technologies, like Xen, a hypervisor that helped treat the vast server farms full of hardware as a single operating system (and to bill for it too, crucially). Thus emerged Amazon Web Services (AWS) comprised of Simple Storage Service (S3) and Elastic Compute (EC). And while Amazon's margins on selling books and other stuff remained minuscule – barely worthy of the firm's valuation – AWS was almost pure profit.[58]

Outside Amazon's data centers, the question of how to access the data it was amassing also became a concern. One of the most recognizable avatars of the internet we could have had was Tim O'Reilly: publisher of the widely used and recognizable handbooks for making the internet we could have had (books famous for their black-and-white cover illustrations of animals totemically assigned to each new technology that required a textbook). O'Reilly urged Jeff Bezos to "develop a series of online tools called application programming interfaces, or APIs, that allowed third parties to easily harvest data about its prices, products, and sales rankings."[59] He was himself frustrated with his own attempts to "scrape" this data, which led him to think about the value of it to him, and to Amazon. He suggested a remarkably naive proto-definition of a platform. Allowing others to build something with this data would benefit everyone: "'Companies need to think not just what they can get for themselves from new technologies but how they can enable others,' he said."[60]

In O'Reilly's mouth, an open API was part of the self-leveling playing field, the end-to-end innovation that made the internet we could have had great. In Amazon's hands, however, it meant something else: rather than building a public good together, inverting the internet meant selling the very insides of the corporation Bezos had created, but for profit. In Amazon's eyes, there was no public, only competitors.

The inversion Amazon effected went mostly unnoticed in the mid-2000s. The nebulous narratives of cloud computing helpfully obscured what Amazon could do: to shift the internet from being composed primarily of a hugely distributed set of computers of all shapes and sizes around the planet – one that tracked the image of autonomous (individual or corporate) persons – to being mostly contained inside data centers, in clouds, accessible as "virtual" servers and commodity storage. As the actual "end-points" of the internet (home PCs acting as servers, corporation-run server banks, government agencies and their mainframes) migrated into these data centers, the data centers (and the systems for accessing them and managing their availability and cost) fell under the control of a small number of large corporations.

Practically speaking, this meant that anyone figuring out the internet we could have had increasingly found it much easier and cheaper to rent the internet from Amazon than to build it themselves. The cloud we watched form was not a cloud at all, it was just Amazon and its ilk frantically re-designing the internet to be a centralized service easily accessed by anyone who needed a "storefront" and who didn't want to invest in their own servers, engineers, support staff, or programmers. Programmers who, in any case, if they weren't already busy making empty promises in a venture capital tailings pond, would shortly be extracted from it like so much molecular gold – to work for Amazon. Monopoly was for suckers. Why bother monopolizing a market for a single good or service when you can control all the markets for all the goods? If Amazon monopolized anything, it was the very ability to evolve the internet we could have had; it called an end to endless possibility.

But this strange centralization of a decentralized system had a paradoxical effect: it actually dramatically *accelerated* the ability of innovators to experiment cheaply and quickly. Even a global financial meltdown in 2008 would not stop the proliferation of internet startup firms, new social media, and much more. Moving fast and breaking things only happened because the sunk capital costs tended toward zero thanks to Amazon Web Services, who were now making money from the innovators who now rented the internet from them. For none other than Google CEO Eric Schmidt, "it was at least two years before he noticed that the founders of seemingly every startup he visited told him they were building their systems atop Amazons servers. 'All of the sudden, it was all Amazon.'"[61]

Google, Microsoft, Apple, and others promptly followed suit. Renting the internet we could have had became the de facto way to engage with it: as a platform. This was Amazon Web Services: built on cloud-based data center computers virtualized and billable by the hour or the gigabit, offloading the responsibility and the difficulty of maintaining one's own "autonomously bounded network" to a corporation providing a commodity. This should have triggered the end of these corporations' special status as the fragile golden egg they claimed to be and the goose that Congress refused to kill. Quite the opposite followed: more protection, less scrutiny, and far greater concentration of power than was ever seen before.

Within years, the internet we could have had disappeared, but no one noticed. A few Morlocks remained behind under the surface struggling to maintain the internet we could have had. First on home computers, then in server closet rackspace, we begged, borrowed, or stole, briefly on Raspberry Pi experiments, but only in the tributaries of the world's largest river, which was slowly eutrophying as it carried us toward the acid ocean of the major platforms.[62]

Bizarrely, the internet became even more singular; rather than an inter-network connecting up lots of diverse networks, it transformed into platforms, and platforms became the focus of investment and critique. It was a change in perception that resulted from the way people used and experienced the internet and its tools and technologies, how they manipulated or hacked them, how that internet was regulated, and how those norms and forms were enforced.[63] Instead of a global brain, it started to look more like a brain trust; instead of a public sphere, it started to look more like a state apparatus run by private corporations; instead of an efficient market, it started to look like an extraction machine. More capitalist than the capitalism we were saving from the capitalists. The internet we could have had become the internet we do have.

In the end, what Amazon did was probably more Hayekian than anyone is willing to give it credit for – even Hayek himself. The idea that the internet could have been a peer-to-peer decentralized market depended on the idea that the internet would not change its spots. Perhaps it was also because of the dream deferred that was net neutrality: the paradox of a technical infrastructure that could remain the same only through the attempted legal and regulatory protection of this Hayekian marvel. Amazon recognized, however, that the internet was just one more tool among many in a larger decentralized world of spontaneous orders at multiple levels: and boy did they nail it.

The dream of a Hayekian peer-to-peer bottom-up decentralized market of independent producers still exists, but it is now technically impossible to imagine the internet as the medium in which it would come to exist. Today one cannot access the internet on one's own terms anymore: one must go through a Content Delivery Network, host your code on GitHub, run a gauntlet of privately owned payment and distribution technologies, all of whom skim off the cream as you head to market with your CBD-infused milk alternative. The poor loser selling that milk is no more than a *memento vivere* of the capitalism that the internet we could have had was supposed to produce: real capitalists control the market itself, the winner who takes more than all.

I can imagine a shorter version of this story: it's just an elaborate told-you-so. That's how capitalism works. Maybe it's just a win for Schumpeter (creative destruction) instead of Hayek.[64] It's just more proof that all that is liquid evaporates into clouds. It's the hegemony of the oligarchs after all. And yet, there is an internet we could have had. And the internet we do have couldn't have come into being without the one we could have had. Literally: Amazon went from what we wanted to emerge to what we now most despise. The internet we could have had is still with us in ways that are not completely apparent to the theories and explanations that are telling us it never existed. Amazon was clearly a creature of the internet we could have had. They emerged from a system designed to be uncontrollable (or "neutral"), only to turn around and find themselves in control of it. But Hayek would have to say, if he were with us: that's spontaneous order for you. Except he would also be forced to admit that spontaneous order could just as easily give rise to something as bad as, or worse than, the fascism to which he sought antidote.

The internet we could have had ushered in a mutation, or a new possibility, within capitalism, in which just being the best product on a market is for suckers.[65] Owning the market is the new goal, and why stop at the market – why not own the very structures that make markets work? Why not own the money, why not own the ledger, why not own supply and demand itself?[66]

Markets, if you believe in them at all, are supposed to make better choices than individuals do. In economics, a core analytic concept is market failure. The virtue of the market is that it tames the vices of individuals – it is a variant of popular sovereignty in the sense of placing limits on individuals and firms, who otherwise might seek to dominate others. The twentieth century saw the invention of the idea of "market failure": that moment when the government (the people) step in to adjust the limits. Amazon's success is *market failure at scale.* It is not simply a question of one particular market failing because of manipulation or corruption, or because of the character of the commodity market or the structure of the labor market: it is a question of a whole lot of markets failing due to the power of one central ~~market~~ player. The market Amazon controls is the market under erasure, there but not there. The internet we could have had was both the bulwark against this and, in a weird way, enabler of Amazon's innovation and the source of the power it has amassed. What it has produced, perhaps, is not just more of the same capitalism, but *two capitalisms,* one feeding off the other. And if there are two, there are probably more.

The internet we could have had – in the era of globalization – was a test of the compatibility of capitalism and democracy. Could they be hybridized, as the end of the Cold War had promised a new world beyond competing systems of communism and capitalism? Instead of a glorious fusion, we've seen a monstrous fission: more than one capitalism, some governing others. Amazon and a handful of other corporations control, design, steer, or incorporate other forms of capitalism – the forms we theorized as made up of a social contract between business and the state, the forms we think of as being subjects of regulation, not authors of it. Amazon, Alibaba, Tencent have all come to see their role as providing the markets, the labor, and the distribution as commodities for all the schmucks playing that other capitalism. Even more, they engage in a form of de facto regulation through standardization and technological coordination (see "If Openness . . ."). These aren't "apps" (not even "super-apps"), these are both profit-seeking actors and also actors who have taken on a very large range of functions classically carried out by governments (and in China, with the immense financial support and legitimacy of the Chinese Communist Party). The term "platform" is not up to the job; the idea that platforms are just "two-sided markets" is a rationalization. It is clear that, whatever these entities have become, they combine elements of both capitalism and governance in forms we have never seen before.[67]

We could turn to an older faith: mutual aid, participatory self-governance, cooperativism, peer-to-peer production. But now it must operate within a second capitalism that encompasses that faith, compiles it into a new binary, and throws away the source code forever.[68] A more participatory democracy holds out the promise of a better world, but what would that look like?

If Participation . . .

Once upon a time there was a line between everyday words and published words.

Everyday words concerned what we would eat for dinner or the things we said in bed; published words were stamped in books, inkjetted in newspapers, beamed on television or the radio. Everyday words were rarely recorded or written down, unless they were cursively written in diaries; published words were serious – rarely did they disappear forever because the point was for published words, official words, serious words, to enter an archive that everyday words emphatically did not. Stasi ingenuity, gossip networks, or surveillance experiments notwithstanding, everyday words were freer and looser because less consequential: no one really cared what we ate for dinner, and few people could gain access to our mortifying and ephemeral pillow talk. This was the way we lived.

Published words, by contrast, could have a lasting effect. By the mid twentieth century, "mass" media defined the line: public speech meant edited, official, profitable, and reputable speech. Published words (whether written in words on a page or spoken words in a recording) were intended to have an effect and to perform, but they were limited on all sides. Modern propaganda is a creature of these limitations: because public speech was not available to everyone, it gave immense strength to those who could access it. But it often failed for the same reason: published words were always out of their control because no one can control what people do with them in private. Published words were hard to produce: they could be submitted to ridiculous restrictions ranging from censorship to copyright to the idiosyncratic delectations of a writerly elite. Only in everyday private speech was it possible to control what words do and mean; public speech is almost by definition out of control. Words were *political* when they were public, but not when they were private.

The internet we could have had objected to this line, tore it down, and created an experiment in letting everyone speak. No restrictions, no censorship, no authorities. Like punk rock – call it here the punk internet – it was totally allergic to the domination of modern social and political institutions that had smothered popular power. But just because everyone could speak, did not mean everyone could be heard.

Thus, the punk internet created new problems, largely of its own making: how would you find someone to listen to? How would you know which people to listen to? How would you know who to trust? Search, filter, and reputation. In a remarkably short period of time, all three questions were answered in a mix of technical and legal forms. Many people on the internet we could have had were occupied with searching, filtering, labeling, licensing, and computing all of the words and images that people contributed. At its core, it was a struggle over the *animation of deliberation*: applying the power of computation, software, and networking to the utopia of free and public deliberation.[1]

The Facebook 'Like' button was a simple solution to all three problems: by expressing approval and aggregating it, everything said could be filtered and sorted, and a new form of reputation emerged. It had the added advantage – for Facebook – that it turned that utopia of free and public deliberation into an enormous database which it controlled almost exclusively; it turned a plan for giving the power of public speech to everyone into the nightmare of making all speech private, and the property of one corporation. With this movement, the internet we could have had became the one we do have.

But Facebook's triumph was not only an ending, it was also the production of something new: a new kind of political subjectivity, one in which deliberation is no longer a qualitative affair, but a quantitative, animated one – "the algorithm." This opaque oracle consecrates speech in new ways: not through argument or logic, not through rhetoric or persuasion, but through visibility and resonance – which is to say, through virality. In doing so, it shifts the meaning of speech away from a concern with words, content, or communication and toward a concern with affect, meaning, and sense. It also created a new possibility: controlling not what was said, but what is *meant*. Sense and meaning were once the possession of the interpreters, and logic and rhetoric were ways of sorting out the sense of sense: the meaning that would come to have an effect in the world. Controlling virality, however, is different: it allows one to skip the steps involved in logic or rhetoric; no deduction and no persuasion are necessary, because the algorithm sets the price – quite literally in a market made of billions of signals calculated instantly, and supporting millions of people who earn a living by it – of that meaning.

§

A distinction between public and private speech is important to the existence of popular sovereignty: that people debate and deliberate gives the people a means to govern in some other way than through violence. It is not just the ability of an individual to speak their mind – that could happen at any time, publicly or privately, with differing consequences for the person. Different definitions of the distinction work: Arendt defined speech and action as taking place in the "space of appearance"; Jürgen Habermas described communicative consensus as *öffentli-*

chkeit; John Stuart Mill offered liberty as the resistance to social conformity. More recently, the distinction remains central to the "agonistic" politics of Jacques Rancière or Chantal Mouffe, where dissensus and conflict are foundational, but still do not descend into the mute violence that represents the disappearance, or bankruptcy, of politics.[2] The problem of "hate speech" haunts this theory at its core (so too, in a similar way, does the theory of "money as speech" in *Citizens United*): if speech can be violence, what happens to politics? But for most such theories, the only place where such speech is legitimate is in "the space of appearance": not in private, not hidden from view. People must be able to appear to each other in some way, so that the meaning of speech, and the action of finding "the right words at the right moment," are legitimate, subject to other words, and other speech, limited by the very plurality and presence of people. That people have and want very different things must be addressed through speech, however that is defined. Thus, it matters where public speech occurs, with what constraints, and even what we mean fundamentally by speech.

The locus of such deliberation is the locus of popular power and so it has always been subject to machinations, innovations, and attempts at control: from the storied coffeehouses that reproduced elite distinctions with the veneer of popular sovereignty, to the profitable and partisan newspapers and broadsheets of the nineteenth century that Mill both contributed to and theorized as sources of liberty (and which Habermas identified with the decline of the public sphere), to the dialectical enlightenment of mass-media radio, film, and TV empires of the twentieth century that even Theodor Adorno and Max Horkheimer's caustic cynicism could not corrode.[3] The constitution of that locus is also subject to problems of design that reflect styles, genders and races, cultures, hierarchies, and rules of debate and discourse.[4] Despite pretensions to becoming public, such spaces of deliberation necessarily exclude some and include others, just not always in the same way every time.

The *value* of public deliberation is also a core part of the story of popular power. Speaking in public became big business long ago, and anxieties about money in politics are probably as old as politics, or money. But the nature of the institutions which relate money to politics changes constantly, and as they change so does the line between public speech – the locus of popular power – and private speech, the internal or intimate deliberations of individuals and groups. The twentieth-century concentration of public speech in a corporate ecology of mass media produced a very specific, large-scale kind of popular power, paradoxically controlled by elites – bouncers at the door to the coffeehouses.

Control over the print media and airwaves is always a story of the attempt to channel and control speech to try to extract profit from it, balanced on the other hand by forms of journalistic ethics, media watchdogs, an ethic of the "right to know," and transparency, or truth in advertising. All this competition resulted in a set of twentieth-century gates that limited public speech to the few who could get through; the bargain in being allowed to enter and speak at scale was to speak differently as a public entity than one would as a private person. The commercial capture of public speech as a source of profit also changed the values associated with private speech.

Everyday speech, private speech, speech and thought that were not in the space of appearance: such speech paradoxically increased in importance as a result of this commercial capture. Mid-twentieth-century economists and social scientists invented a kind of sensor for everyday speech: public opinion polling.[5] Public opinion polling is the aggregation, weirdly, of *private* opinions: opinions held by individuals, representing, supposedly, their most cherished beliefs and desires for the future. Economists and pollsters told themselves bedtime stories about rational individuals, utility curves, and methodological individualism which would determine what men think and do. The aggregation of these data fed sophisticated tools, some designed for analyzing economics, and some for activities like voting, or expressing other political preferences. Surveys, polls, periodic election results, demographics, "revealed preferences": all ways of quantifying some version of "interest" (by analogy with "utility" in economics), which, when summed, could stand in for the general will, measured, monitored, aggregated, and displayed on the nightly news and in morning newspapers.

Public opinion polling was critiqued because it does not accurately reflect popular power, or it misses the qualitative nature of democratic deliberation, turning it instead into a kind of proxy voting. But it also has a positive effect: it enters into public speech by reflecting people's interests back to them. Public opinion becomes, recursively, an object of debate and deliberation itself. Polls and surveys have long been ways of speaking about ideas at larger scales, with abstractions whose rhetorical force has long since become central to nationhood.[6] Nonetheless, these tools are confined to a certain kind of bloodless, scientific affect: the truth, submitted to statistical and probabilistic quality assurance mechanisms, packaged in pie charts and graphs, approval and disapproval ratings. The industry of polling that arose in the shadow of mass media became enormous, profitable, and, despite all its best intentions to neutrality, central to the political subjectivity of twentieth-century democratic debate.

Everyone's daylight hours were saturated by the organized and profitable information that public opinion provided. We learned to reflect upon the quantified interests displayed to us as we visited the news-stand in the morning, listened to the radio at midday, and watched televised news at night. We conversed constantly, fugitively, in the interstices; the conflictual relations we experienced in everyday speech rarely felt powerful; most of us never felt heard, visible.

In the twentieth century, it seemed that everyday public deliberation receded farther and farther into the distance;

political subjectivity was confined to expressing opinions and voting in elections, two sides of a tarnished coin. It was a world in which having an opinion was a private matter; only in rare instances could it become public through this peculiar form of aggregation.

§

The punk internet would erase the line between everyday speech and public speech, just for a moment: it would open up deliberation again, make all speech public in a space of appearance. Not only did it open a possibility that people had long since come to think of as impossible in the face of mass media and global social and political institutions, it also created a new kind of intimate space as a result. It is not an accident that people immediately fastened on to the trouble it made with *reality* (virtual reality, online and off, life and second life, etc.). In the precise sense that Arendt gave this word, this reality was about being together in a space of appearance, something that in the twentieth century had faded from everyday experience, replaced instead by, perhaps, the society of the spectacle.[7]

Thus, the internet we could have had was a startling experiment. All of a sudden, everyone could speak, everyone was together: everyone appeared, potentially, to everyone else who was there. The very idea that it would include *everyone* introduced the same problem that the mass media faced, at an even larger, more threatening scale. It was a problem of sorting out who could speak, who could be trusted, who would be heard. But the solution would be different this time around. Rather than quantifying and displaying interests, it would *animate deliberation* (the expression, contestation, and interpretation of interests).

§

The punk internet emerged with punk itself: in among the counter-cultural appropriations of various media technologies, from Portapak video cameras to Xeroxed zines to lo-fi music production. It emerged along with feminism, whose 1970s clarion call was to recognize that the personal is political. The goal was to tear down this line so that everyone's words, images, or sounds could be published words, images, or sounds. No gate-keepers; destroy the elitism that smothered private words and prevented them from performing on the grand public stage. Smash the patriarchy, learn three chords, steal this book, let everyone up on stage.

The punk internet was one version of the internet we could have had, and it concerns primarily the process by which speech and action were transformed by the internet. There were, of course, "cyberpunks" – more a literary genre than an actual form of life – from William Gibson's *Neuromancer* ("The sky above the port was the color of television, tuned to a dead channel") to Neal Stephenson's *Snowcrash* ("The Deliverator belongs to an elite order, a hallowed subcategory"). Cyberpunk established the punk credibility of the internet both thematically (Gibson wrote his novel on a typewriter) and by reference to the insider knowledge and terminology of the 'net (Stephenson was "highly online" before it was a thing).[8] There were also the "cypherpunks," an avowed movement of cryptography obsessed anarchist thinkers.[9] And more generally there was a large and chaotic hacker underground.[10]

Above all, the punk internet tapped into concerns about freedom of speech and its central importance to political self-governance in a democracy. This included epic debates, government reforms, and police action related to copyright and patent. But it also included an unlikely array of people interested in tearing down lines and borders of different sorts, from an almost predictable Grateful Dead lyricist declaring the independence of cyberspace (John Perry Barlow), to a radical revolutionary fomenting online activism against the Mexican state and neoliberalism generally (Subcomandante Marcos of the Zapatistas), to an arch conservative neoclassical economist in favor of loosening up IP law (Milton Friedman), to a Gen-X Riot Grrrl intent on remaking 1990s pop culture (Jamie Levy), to net.time pranksters and art-thugs (Jodi, eToy, RTMark).[11]

The making of the punk internet merged into an exercise in remaking democratic institutions of speech, without naming it as such. It was about free speech, public speech, about the ability to form and state opinions, listen to other people, form relations with and be with other people, without any kind of representation. On the one hand, it was a nostalgic techno-town hall (direct democracy), and on the other, an unknown next-generation democracy.

§

The only people opposed to the punk internet, said the punk internet, were the suits who owned the media.

No one could break the concentrated media machine; one had to either play the game or find a way to work around it. No one could reform it, and joining it was selling out. Mass media corporations and their allies had built Maginot lines of intellectual property and licensing; they fortified national standards for frequency modulation and color (colour?) encoding; they dispensed the ISBNs and they staffed the ASCAPs and BMIs. They had developed elaborate, effective, and profitable means by which they controlled the words that people could use, the music they could play, the patents they could license, or the brands and images they could appropriate. Resistance was futile.

The punk utopia of the internet escaped this control because anyone could make a website.[12] Or a mailing list, or a piece of software, or a new protocol, or an IRC channel, or a game, or a gathering – online again, offline again. More significant, perhaps, was that a website that such an anyone made could be made publicly available without another anyone's permission. Like punk rock, it took a little bit of effort; one had to have some money, some time, know where to buy a guitar and figure out how to plug it in to the amp. A certain technical comfort zone attracted a certain type of person. "Anyone" is not "everyone," of course, but the punk internet was not confined to a type of person or a particular social group. When it appeared, it appealed to more than just the punks: sports fans, gamers, artists, shysters with glue-on hair products, filmmakers, and bankers; people in India, China, Latin America, Eastern Europe, as much as the US or Europe. It is important to make this distinction clear: whatever barriers there were to people making use of it, they were not those of inclusion/exclusion. At the outset, the barriers were being torn down, not erected.

Making a website was more than writing a punk song without permission: what attracted people was the idea that it could be changed, updated, and, especially, *linked to other websites*, changing the very meaning of finality. It was not a song or a novel – it was a temporal relation to the makers of songs and novels as much as the songs and novels themselves. It was as if your readers were present like audience members at a gig, but audience members who came up onstage and performed alongside you. Was this a public space, a space of publication, or was it a private space, a jam-session with friends, a community who cared for one another? With time, the punk utopia of the internet would connect more and more such spaces. Gigs would find gigs, far away and close up. Instead of a road trip, we started to live in all the cities at once. Instead of a postcard about a performance, the show showed up in the mail and performed for you.

Arendt's distinction concerning labor, work, and action could help here: the novel, the song, the film are all *works* insofar as they are objects of a craft intended to outlive their maker. One looks to these objects in their finality and endurance, whether for edification or affect, beauty or sublimity. But the act of creating a website (or speaking online, or making and updating software) was never finished – it was more like what she referred to as *speech and action* – it was constituted within a web of relations concerning that which is common to these relations (not common to all, to "humans" generally, but to those who appear to each other together). Speech and action happen together and "they leave behind no such results and end products." Being together in this public way (in a space of appearance) was a way of getting around the existing institutions that managed and controlled public speech.[13] But it was also different from speech by this definition: being together on the internet did in fact leave behind results and end products, and this is part of the story of what has happened to politics on the internet we now have.

The accessibility and flexibility of the punk internet was central to its power: people everywhere brought curiosity to bear on it. What could it do, what were its limits? Was it a medium like film or radio, or more like poetry? What would it do to film or poetry? That curiosity was enabled at a very large scale: the internet was, by design, extensible, and so people everywhere began to extend it. Extensibility invited the curious to learn the internet from the inside out; "view source" was right there below "open" and "save"; websites hosted not just text but code that could change what the internet could do, and it was being shared freely, not just sold for a price.

§

Questions emerged: how can you find the information or the people you want to find? How can you filter out what you don't want to see? How can you trust the connections you do make? The answers to these questions were never just answers: they were also about *making solutions* – building websites; coding; designing protocols, standards, or proposals; sharing that code, refining it, teaching each other, and also, yes, trying to make money doing it. Debates about "the public domain" were common; intellectual property law (previously an arcane and low-status field of law) suddenly became, as Jamie Boyle would point out, the very skeletal structure of the information revolution in progress.[14] Free software emerged at the center of all of this (see "If Openness . . ."), with its licenses and its debates about the nature of speech (was code speech?[15]), the nature of constraint (could legal restrictions be technically implemented?[16]), and the definition of freedom (negative, positive, etc.[17]).

"Making a website" was much more: it was about diagnosing problems in the real world, proposing real alternatives to them, and then living in those alternatives, trying them out for real. These worlds opened up: rapid and unconstrained experimentation with writing, speaking, making, recording, engaging, circulating and remixing, publishing, reviewing, commenting. Every such experiment was an invitation to join and test it; every disagreement was an encouragement to try out the alternative, not simply dispute it.

In a remarkably brief period, from roughly 1995 to 2005, the punk internet created forms of software sharing, licenses, art works, manifestos, protocols, standards, websites, web-logs, blogs, social software, resource description frameworks, microblogs, archives, encyclopedias, peer-to-peer systems, bulletin boards, games, spaces, educational resources, industry innovation experiments, remix cultures, and mundane business concerns.

Eventually this utopia was either no longer punk, or it was the apotheosis of punk. But it was clearly a triumph over the line between public and private speech. And it worked. It let everyone in, even the people no one wanted to hear from and especially the people who would not stop talking.

Thus it was that, by 2004, the punk internet could show up in an unlikely place: a Harvard dorm room and an IBM PC hosting a directory of Harvard students and a small piece of software that allowed a viewer to rank each of them as "hot or not."[18]

§

In the punk utopia of the internet, it soon became clear that just because everyone could now speak, it did not follow that everyone would be heard; it did not follow that everyone ought to be heard, and it did not follow that everyone could be trusted. As an experiment in animating deliberation, the internet we could have had engaged in a process of solving these problems, collectively and publicly. But what started out as a punk utopia developed instead into a set of platforms with a nightmarish level of control over speech: a world in which there would eventually be almost no public speech, only private speech, polarized, hate-filled forms of deliberation. How did this happen?

The first problem that the punk internet encountered was that it had no map. It seemed to come out of nowhere and often lead nowhere. No obvious channels to change, no table of contents. Because there were no gatekeepers, there were no entry gates. And so the earliest experiments were directed at this mapping: "best of" lists, links and web-rings, primitive forms of search, "web crawlers" proliferated quickly. The emergence of "search" as a ubiquitous relation to information is directly related to the fact that the punk internet let everyone in, without needing permission: it grew way too fast and was designed to be controlled by no single entity. But search was just the first problem.

The second problem of the punk internet was how to filter it, how to hear what you wanted to and not what you didn't. Spam was an early warning sign.[19] Spam is the opposite of the public discussions that formed online – it represents the intrusion of something unwanted into a desired space, but, more importantly, it takes up space that could be filled by others. Spam instantly heralds a dystopia, in which the newly public speech, ungated and uncensored, is drowned in the speech of another, whose concerns (green card lotteries, ED drugs, and get-rich-quick schemes) immediately let the air out of the newly powerful public speech.

People joined the internet we could have had as individuals; they often joined in multiple ways simultaneously as individuals. Practically speaking, people did this pseudonymously. User names, now so ubiquitous, were the site of imagining and projecting personhood, real or aspirational. It was not quite anonymity – people brought their personalities and their ghosts with them – but it lacked the fullness of personhood; personhood was limited by this new medium. As people came together pseudonymously, they formed a collection of pseudonymous people: people connected only by their speech and action in this domain, and they immediately started to experiment with these forms.[20]

In a world without gatekeepers, who would keep the unwanted out? For a certain radical version of free speech that emerged with the punk internet, the answer was: no one. Anyone could speak because that is the whole point of free speech – not what they say or how they say it, but their right to do so. For the most austere, this included "commercial" speech as well as ostensibly political speech, and so it was not always obvious who considered which speech to be, for instance, spam as opposed to a political opinion. For the most extreme, it was the origin of sites like 4chan, experiments in the most negative form of negative liberty the world had yet seen. Ecumenism breeds extremism: the more the definition of rights and inclusion expanded, the more extreme the transgression would become.

And since the punk internet was an experiment of living in a self-made democracy, the problem was not primarily about whether the actually existing governments, on the outside, would endeavor to restrict speech on the internet (the 1996 Communications Decency Act being the most consequential of such efforts). Rather, it was first a question of whether anyone on the internet could restrict what they themselves were saying. Because this was a public space of appearance, a site of radical popular sovereignty appearing suddenly, and possibly – hopefully – as a successor to representative democracies, it was a question of whether the people can restrict their own free speech. Who would do so and how?[21]

Pragmatically speaking, the answer was: those with root access. The moderators of mailing lists, the sysadmins, the people with the private keys. These petty tyrants appeared everywhere in the internet we could have had, exercising arbitrary authority over everything from who could have an account to the style of speaking allowed on mailing lists to the acceptable uses of software: the punk internet may have been an experiment in freedom but it would also be an experiment in arbitrary rule.

Such forms of power were widely seen as illegitimate; but, more importantly, they were also seen as self-limiting. If a website or a mailing list restricted something, then make a different list or website; go elsewhere. Extensibility was its own solution to arbitrary limits, whether autocratic pettiness or benevolent dictatorship. In some cases, the process was to reproduce internally the institutions of democracy and representation, and so proliferated a wide range of online governance systems, ranging from the autocratic to the robustly participatory democracy.[22]

Anonymity and pseudonymity were in tension here: on the one hand, there were those who believed strongly that speech and identity should be disconnected from each other. People should not be accorded more power or authority because of a quality like history, race, credentials, status, class, biological sex – but only on the grounds of their speech and ideas. But on the other, pseudonymity was an opportunity to try out new types of identities, new forms of social relations, new persons. Combined with the internet we could have had, it was a palpable space of self-expression that could be compared only to nightclubs of a certain sort, perhaps, or underground communities all too often labeled deviant; see also basically any cyberpunk story.

Thus, the internet we could have had was a technical experiment with radical free speech; only the content of speech mattered, not its speaker, their history or expertise, or the context. But it was also an experiment with new ways of being together, of appearing to one another. Few people trusted such a radically speakerless discourse. Stable names were necessary, even if disconnected from real bodies. Pseudonymity was compatible with *reputation*; radical anonymity risked irresponsibility. A new definition of reputation might conceivably determine who to listen to, how to sort signal from noise. Reputation's power became attention, which was theorized early on as the scarce resource in a world getting noisier by the day, because the old wall between public and private speech had been breached.[23] Hence, squaring the value of speech and the reputation of the speaker became the hardest of the problems on the punk internet.

How to have freedom of speech and reputation at the same time? This was the third problem of the internet we could have had. The basic form in which this experiment took place was, at first, small-scale: mailing lists and forums and newsgroups and other sites of association. Rules and forms of etiquette were innovated, moderators were elected, forms of power and restrictions on that power were proposed. The punk internet had torn down old walls, however, so was this just the reinstallation of the *ancien régime*? A new wall between private and public? The nigh end was declared whenever the petty tyrants of Slashdot or nettime acted too fast, without impunity, as "benevolent dictators" too quick to ban or too laissez-faire to maintain comity.[24] The "flame war" emerged as a distinct discursive style of engagement.

On this small scale, participants often acted in terms familiar to any reader of John Stuart Mill.[25] Speech and action fills the world with ideas and ways of living, some good and some bad; more speech sorts those ideas out, counters them, tests them. The dangerous and threatening ones are neutered as necessary, or rejected; the good ones are reaffirmed and the wheat is separated from the chaff. Good ideas become better by repetition, and so do the arguments against bad ideas.

The qualitative magic of deliberation was, on its own terms, worth preserving, but the internet we could have had was getting larger by the day. Sometimes participants in these debates would say to those who objected: go elsewhere, start your own list, fork your project, make your own version; the internet can sustain an infinite number of collectives, this is its magic, it is decentralized and extensible, new worlds are endlessly possible!

Moderation became a topic of explicit, recursive discussion; people proposed metrics for upvoting and downvoting, techniques debated ad nauseam on sites like Slashdot or Kuro5hin. A different solution emerged slowly and through the very fact of the medium itself: to apply the quantitative power of software to the qualitative magic of deliberation. It would not be a question of moderation, not so much the question of saying yes or no to a particular voice, but instead the challenge of finding a way to combine many voices, like the outcome of a vote.

§

The three problems of the punk internet – search, filter, trust – were also efforts to define the internet we could have had. Would it be, as questions of searching and filtering seemed to imply, a giant brain, or a library, or a public space in which ideas and opinions circulated freely? Or would it be, as the problem of trust seemed to imply, a new version of a collective, a new kind of relating with other people, like a society, or a market, or a community, but better?

Both definitions were medium-specific. First, the problems were wrapped up with the fact that the internet we could have had produced, at a new scale, all kinds of *works* intended to circulate without gate-keepers: writings, software, music, video, etc. Second, the internet we could have had also seemed to be a kind of participation (speech and action) in the way it brought people *together*. Because it was still only the internet we could have had, no one knew for sure which of these things it would become, or if it could be both.

Some early denizens saw the internet we could have had as a resource for a prior togetherness they were not ready to leave behind: it was like a community library or a bookstore. It was the memex of Vannevar Bush, for instance; or it was a recapitulation of the library from the arts of memory to the card catalog, in digital form. Information and knowledge were inputs to public deliberation (or for that matter, innovation) and not the deliberation (or innovation) itself, which happened elsewhere. It was merely a resource for participation, not the act of participation itself.

But for others, the internet we could have had was the sudden (re)discovery of a space for relating to others – being together that was inaccessible in the existing world of mass media. It was a locus of speech and action, potentially of new beginnings, but one that occurred via the mediums of writing, reading, inventing new sign systems, or making new tools for manipulating sign systems (software).

Taken together, the problems of searching, filtering, and trusting wavered back and forth between information and people: Were these problems of searching for information, or for people? Filtering out speech, or excluding people? Trusting a source, or trusting a person? Whether information or people, the challenges were also confounded by scale: more and more people were joining the internet all the time and, as such, more and more information was being generated all the time. How would the internet we could have had manage all the world's knowledge? Collect and organize everything, which was getting bigger every day? Similarly, how would people find and connect with each other, as the number of people grew and the spaces in which they gathered proliferated?

The fact that the internet we could have had offered the very tools one needed in order to address these problems lent the "recursive public" its force: we can use the same medium to solve the problem which the medium is encountering.[26] We can build technologies that will automate the processes of finding things, filtering out things, and knowing what is valuable.

Thus, a beginning. The punk utopia thus gave birth to another utopia inside it: a homuncular utopia in which perfection consisted not simply in letting everyone speak, but in finding and filtering all the world's knowledge, making it easy to locate for any purpose, a public library but massive, evolutionary, and emergent, a "noosphere": *the semantic web*. At the same time, this homuncular utopia would be a space in which human thought and desire were clear, collective, and conscious, but under no single person or nation or corporation's control. It would be the will of the people, by the numbers. A digital *conscience collective* – a decentralized, addressable, and searchable General Will. Such a little utopia within a utopia is easily missed when one looks back on the history of the internet with Silicon Valley-colored glasses on; the very fact that this tiny utopia had such a wide range of practical, technological, and legal outcomes raises the question of the book: How did the internet we could have had so easily become the internet we do have?

Of course, this homuncular utopia was built on entirely mundane technologies: programming languages, distributed operating systems, browsers and webservers, protocols and standards. The power of free and open source software (see "If Openness . . .") meant that such tools could spread rapidly, improvements contributed by one person becoming the property of everyone. A working tool required its own array of new relations, mailing-lists and meet-ups; new standards and organizations formed. The "utopia of work" was the way to extend the system: coding improvements and repurposing thousands or millions of old PCs into a rickety internet that drove the very enthusiasm and agitation of the punk internet right up until Amazon inverted it (see "If Decentralization . . .").

The homuncular utopia was also built on the backs of small standards organizations, non-profits, associations and foundations designed to promote democracy – but was unable in the end to defend democracy. Such were the agoras of most consequence: spaces where technically steeped denizens of the internet designed and hammered out standards and tools, slowly and deliberately, in forums designed for qualitative speech to emerge – mailing lists, in-person meetings, and conferences; spaces designed to allow debates to come to a conclusion, and decision-making to achieve consensus.

The three problems of the punk internet came together just at this moment, right at the heart of the dotcom boom, and the political economy of the internet we could have had started to become clear. The "browser wars" of the late 1990s demonstrated the power of open source and free software and created the moral high ground on which much of the ensuing story unfolds. There was a famous attempt by Microsoft to "embrace and extend" the core protocols of the web in its Internet Explorer browser. They pursued a devious course, implementing all of the core protocols, and a few bells and whistles of their own, in an attempt to capture the already-famous "network effects." All this was leaked and much discussed as the malfeasance of a greedy, controlling corporation.[27] David to Microsoft's Goliath was the decision by Netscape to make the source code to Netscape openly available, a first for a large internet corporation and a win for the punk internet at the time.[28] From this moment onward, the "open" internet, the "end to end" principle, net neutrality, and "the semantic web" emerged as the inevitable future of the internet we could have had.[29]

The "semantic web" was the vital force of the homuncular utopia: it was really the utopianists' utopia, the perfect world where map and territory converge, the solution to all the problems of knowledge and representation; talk of "ontologies" is free and easy in this utopia, and no one ever fails to say what they mean. To many of its partisans in the late 1990s, it appeared to be just around the corner.

The semantic web appeared primarily in the scholarly circles that made up information and library science in the 1980s and 1990s. Over the years there had been significant effort to model information's meaning in the wake of an information theory that declared that it didn't matter. Shannon's information theory was the organizing model for the mathematics of information, but it was rigorously indifferent to the subjective source of meaning. If there was meaning, it was encoded mathematically in the communication itself. Such a theory was pragmatically unsatisfying for many scholars interested in creating digital libraries, information ontologies, or in helping to make possible tools for finding and making meaningful connections. A true semantic web would not only mark-up text for humans to read, it would mark it up for any-computer-whatsoever (a Turing machine) to read and process. Semantic labeling was necessary for this to happen: you couldn't just say something, you also had to say what it meant. Predictably, the logic and engineering of a true semantic web is complex, but ideally slightly less complex than language and meaning itself.

Tim Berners-Lee gets all the credit, but he was far from alone, or all that ambitious, when it comes to the WWW. In part, because his version (HTML) of an older markup language (SGML) was simple and easy to use, it stuck to the internet we could have had and began generating vast numbers of websites, upon which people were clicking in increasing numbers and for diverse purposes. HTML made it easy to make a website look like a *normal document*; it didn't have to be the WarGames-style green screen, or the data-entry menu of 1-key commands, it could just look like regular text, ideally in Comic sans font, with an animated gif in one corner – GeoCities.

What HTML did not do was require that information be semantically labeled. An army of information-obsessed punks on the internet picked up on this theoretical hole. Wasn't it a shame that all this information being added to the internet was *unlabeled*, and by being unlabeled, was uncomputable, qualitative: only good for humans? Only human-readable. A proper global brain would have to be machine-readable to become the noosphere of the punk utopian internet imaginary. HTML and its ilk were good at labeling information for *presentation*: how it would be displayed on a screen to users, left, right, bold, italic, and so on. But the semantic web would extend this idea, covering not only how information was displayed, but what it *meant*: its ontology. A picture of an apple would not just be labeled with an "img" tag and some code telling the screen where to put it, it would also contain a "fruit" tag or a "red" tag" or a "food" tag.

The idea of the semantic web was absurd though. To achieve it, the denizens of the punk internet would have had to not only say what they wanted to say, but also to say what they meant. To say that "Joe Strummer is a fan of blues" required not just writing the sentence, but also labeling "Joe Strummer" as a "punk musician," "blues" as a "type of music," and "fanOf" as a particular relation between the two. No one had time for this at any level. The fact that the very idea had a huge number of enthusiastic supporters and generated a complex set of standards and protocols boggles the mind. It was all but set to disappear completely, sometime in the mid-2000s, despite its fandom and an extensive amount of work by people around Tim Berners-Lee and the World Wide Web Consortium, all singing the blues about the non-adoption of the semantic web.

§

There are the things that people thought that the internet we could have had might have become; and then there are the surprising things they discovered when they started to explore it. One of the more surprising things – so mundane and ubiquitous as to be overlooked – were the *logs*.[30]

All around the world, at different times and different places, but mostly around the late 1990s, people on the internet we could have had discovered that online interaction came with logs. They were not logs of the content of speech or the software source code, they were logs of internet usage: the connections made whenever someone "spoke." A website was, on the surface, a changing record or words and images; one could keep track of such changes the way software engineers kept track of code. No, the logs people found interesting were the ones that recorded the *action* of connecting: visiting a website, following a link, connecting to a port, and downloading something.

Logs and logs of data about interaction, useless on their own but illustrative of a possibility: logs represent something, but what? What does it mean when a browser visits a link in a page? Curiosity, desire, trust, value? What does it mean when a browser returns repeatedly to a website: joy, depth, or maybe the website is set as the default?

In some ways there was a simple and unintended design decision at the origin of all this: operating systems like Linux and webservers like Apache, as well as the various forms of software that ran on or through them, by design generated detailed logs of interaction. Such logs were largely created by programmers in order to understand when errors happened. These logs were consulted only when something went wrong, and archived and regularly deleted as the conditional processing churn of software passed daily into the irrelevant distance. "/dev/null" was their usual destination, as long as everything was working. A record of what a computer does every second of its operation is of little interest to anyone other than the mechanic. But, flash on the fact that it is also a record of people using those computers to do valuable things . . . well, then all of a sudden you have a strange and powerful registry of everything that occurred, between people and their screens, between clicks and software, across hardware, networks, protocols, and standards.

Logs were logs of participation, an archive of speech and action – not of the content of that speech, but of its effects. Logs suggested that there is meaning in the actions of what people do, not just in the content of what they produce or read. And what people do is connect. The meaning of that connection was obscure, to be sure: it had to be attributed. A log of a visit to a website could be interpreted as a desire for something on that website, even though no one other than the visitor could know for sure. In the absence of more information about who that visitor was, it was thin evidence. But the logs were there, and importantly, they were *already digital*, already computer-readable. With a bit of coding, such logs could be turned into data – they could display a graph of connections, they could be used to visualize the internet and all of its relations, and they were. With a bit of processing, one could *animate* the temporal process of speech and action – deliberation – on the internet.

From roughly 1996 onwards, the existence of logs drove the idea of third-party tracking of all sorts.[31] The tension between anonymity and pseudonymity had found a new twist: being together with others on the internet left traces. That interpretive problem troubled people: how could you know why any given person visited a website, or what they found valuable? How, moreover, could you know if the people visiting a website had anything in common with the other people visiting the same website: what did all the logs and data and computations mean?

It must have become obvious to many people around the same time that the trio of problems facing the internet we could have had – finding things, filtering out things, and trusting people – might all be connected somehow. Perhaps it was obvious to many people that something such as the Facebook

Like button would have to be invented. In the end, of course, someone invented it.[32]

. . . Then Facebook

The Facebook Like button was a clever way to solve all three problems at once: the problem of searching and finding, the problem of filtering out noise, and the problem of knowing who is to trust.

The first, and arguably most important, step was that Facebook had decided to try to make the web safe for people by making it easy to use: people could safely log in and see only what they wanted to see in Facebook, connect only to the people they wanted to connect to, accept or reject such requests, all the while entering their own information – name, location, status, hobbies, political beliefs, and, by inference, social location and relations. Slowly, through the participatory action of its users, Facebook's ecosystem became more "semantically rich" because its users enriched the space, just as people could do on the internet generally. Facebook filled with the "works" of people documenting and describing their everyday: from what they ate for breakfast to how they joined the protest and back again.

The Like button made the "work" into a kind of "action": it transformed what looked like a diary into an interaction among people, but of a new kind. Clicking that Like button provided an explicit indication of a relationship between the person doing the clicking and the thing clicked upon. Clicking produced not just logs invisible to the clicker, but an explicit set of statements: a numerical increase (+1 Like), a new statement that "X liked this post by Y," and a message to the Likee informing them of being so Liked. All this extra communication was strictly speaking unnecessary; it was also revolutionary. Because it was called "Like" there would be no disputing what it meant (and the debates about, and advocacy for, a "dislike button" occupied many people for years).[33] A Like, tied to a person doing the Liking, and a thing being Liked (often a statement or post by some other user): this was the semantic web, in possibly its simplest form.

What Facebook was creating was a version of a "social graph," understood as the web of all relations that exist at any given time. In abstract terms, people on the internet we could have had theorized regularly about what constituted such a thing; it ranged from the more arcane network theories of people like Harrison White to the popular Stanley Milgram experiment that launched the idea of "six degrees of separation."[34] In concrete terms, such a graph had to be built to become real; otherwise it remained a model of social interaction only vaguely visible to the social scientists and psychologists attempting to reveal it. The Like button succeeded in automating a version of interaction not only by registering the relations, but in a very simple form (one-dimensional even) by registering the content of those relations as well: X liked the post by Y; X likes Y product.

Facebook glorified this version of the semantic web by creating something called the Open Graph Protocol. OGP was a vastly simplified semantic description framework: it had none of the pretensions of the universal languages imagined by academics and digital librarians of the semantic web, it was a simple world of goods and services, liking and buying, advocating and recommending. Facebook's ontology had posts, products, an image, a name, news stories, maybe people or keywords, and not much more. Each time a user Liked something, the Like, the user's identity, and the object Liked were tethered together in a database. By making the activity of Liking into a quantitative and not just a qualitative relation (different from and far easier than posting a comment to a website or blog, or engaging in debate on a mailing list), Facebook both eased the process of deliberating and made it numerical. Liking something could be a signal for others to read, and maybe they too would also Like that thing. Nonetheless, no one knew quite what Liking would do. Writing in 2013, legal scholar Leigh Ellen Gray could still assert of Liking that "uncertainty persists as to what Liking something actually connotes as well as what the practical effects of the action are."[35] It would take a moment for people to realize what was happening: it dawned much sooner on some than on others.[36]

Facebook put Like buttons on everything in Facebook: post, product, or person. At some level, this is an uncontroversial thing to do – but what Facebook did next will shock you.[37]

§

Despite having a billion users, Facebook was aware that these users often visited other parts of the internet. Facebook didn't know what they Liked there. This left a hole in the daily diary. To remedy this, Facebook needed two things: a way to make sure that a Facebook user remained a Facebook user even when visiting other sites or services, and the assistance of people in those other sites, over which they did not have authority, in making the Like button available.

The first, Facebook Connect, was one of a series of technical solutions (the general field of single sign-on, or identity authorization applications) designed to fix a perceived flaw in the internet: the fact that anyone could be there, with no authentication of their real identity. What was, on the internet we could have had, a virtue (Tear down that wall! Explore your personhood pseudonymously!) had become a barrier or threat because of a perceived lack of protection. Facebook had recognized and pioneered this anti-anonymity approach (controversially) by requiring users to use real names; Facebook Connect extended this by allowing Facebook users to sign into any number of other websites, while remaining tethered to Facebook's network and its logs.

Second, once logged in via Facebook, it was a question of making things on those other sites – products, news stories, blogs, videos – Likeable. Facebook thus provided, for free, a cut-and-paste javascript program that made it trivial to embed a Like button on any web page anywhere. Developers at other sites did so happily once it became clear that when one Facebook user clicked the button, some large number of that user's friends got a notification. The button clicks, Facebook logs, traffic moves.

Facebook thus found a clever way to turn openness into a possession. The Open Graph Protocol became a sluice gate through which the once pseudonymous speech of the space of appearance flooded directly into Facebook's databases. Facebook used the actions of users to describe their own relations and desires, and collected the logs of all of this relating, in Facebook and on the open internet, as a database they controlled. Facebook's value was now in the giant database that the Like button created for them, by automatically logging every action taken between a person and any semantically nameable entity on the web. As Halpin puts it: "This was a radically successful coup to control not only the social graph but the identities of people across the entire Web. What Facebook began was nothing less than the primitive accumulation of social relationships; the OGP in combination with Facebook's profile data reified social relationships and their related affects to create a giant database for advertising purposes."[38]

Page-views, clicks, Likes, links, hovers, swipes, pauses, and lingerings, were collected into tiny avatars of human desire, affect, and intention that could be assumed, consumed, and recirculated. The Like button – eventually a whole "Like economy" – emerged out of this inversion of the openness of the web into a giant database controlled by Facebook.[39]

Liking, of course, is not the only thing of its kind, but it was one of the first. Variations on the theme followed: retweeting, reblogging, tracking views, downloads, etc. A world of "sharing" emerged with its own icons. Facebook's Like button is not the only culprit in the transition from the internet we could have had to the one we do. One could tell the same story here by focusing on Google. When Larry Page and Sergei Brin created and demonstrated (and patented[40]) the page-rank algorithm in 1998, it addressed that first problem of finding things, and it was part of the punk internet. Not because it was a good idea, but because it was an idea that worked and that people could imagine using, modifying, and extending. Google's choice to keep that system under patent was a sign of things to come for the internet we could have had, but it wasn't until much later that Google would recognize the same thing Facebook did: that with each click on each link of a search result, a user was not just finding something and making a choice, but expressing something like a desire, an opinion, or a meaning. Google exploited that data ruthlessly from the mid-2000s onwards.[41] Don't be Evil was their early mantra: "don't break the internet we could have had" would have been a plausible translation. But they did, and we joined in by doing something that seemed so quotidian and inconsequential: searching, clicking, Liking, and retweeting our way into the internet we do have.

It is worth noting that it is just such details that Zuboff uses to construct her theory of surveillance capitalism; but she sees in these technical decisions the deliberate construction of a system of behavioral control that leads to total domination in "the hive." No doubt this is as good an explanation as any of how Facebook enriched itself and broke the dreams of the internet we could have had. But it was also something more: a new beginning, the creation of something new which is not under the absolute control of the corporations who stumbled upon it, even if they were the first to capitalize on it.

The coup at stake here is not just the massive wealth extraction machine but a transformation of language and meaning itself, at least insofar as language and meaning take place on digital networks and social media interfaces. The act of Liking is not just a vote; it also associates those doing the Liking with each other in a graph. It is a graph that ostensibly simply represents offline connections: who your friends are. But simply step two degrees of separation away and the graph connects you to all kinds of strangers in collectives you did not join or make. A social graph enables, and almost demands, for anyone with access to it, the writing of algorithms that process this graph in countless ways.

The seduction of this data-filled social graph was undeniable. If it were simply a numerical tally, it would be little different than the polls and surveys that made up twentieth-century mass media and its peculiar system for gate-keeping public deliberation. "Sharing," however, does more: it generates a gigantic data object that can be processed and dynamically adjusted in as many ways as one can hire computer scientists to imagine, tuned to produce effects, black boxed. No one who was thinking about the internet we could have had was thinking about reproducing the world of polls and graphs: what held out promise was the computational power of distributed computing, the ability to *animate* the graph and see where it would take us.

Facebook appropriated the space of appearance and revved it up at the same time. By channeling all that participation on the open internet into a database which it controlled, Facebook created not only an advertising revenue system, it generated the unpredictable forms of speech and action we have today. It rendered this collective experiment – the internet we could have had – inaccessible other than through participation in Facebook's own experiments with the model-world it had created. Facebook never lied about this: they wanted to make the web easier for people to use, to make it easier to find friends and content, and easier to log in once and access sites and content everywhere. Like Amazon, they simply wanted to *be* the internet. They wanted a map that converged with the territory; controlling the map was no longer about controlling information about the world, it was about controlling the territory itself. What was before a representation of the world was transfigured into a tool for performing that world, for manipulating that world. And it was all "open."

Because the line between public and private speech had been melted down by the punk utopia of the internet, Facebook's databases were filled with every known form of interaction: from the most intimate to the most formal and public. The logs of our daily activities looked less like a diary and more like the abundant, detailed evidence of the most private of interactions and the most thorough account of our public and social interactions.

It was a social scientist's dream; and it was an advertising company's Nirvana. Even before Snowden revealed the spying of the National Security Agency, Facebook was pioneering a version of it in the open. Logs turned into data, and data was turned into profiles, and profiles were standardized and verified and shared and, eventually, bought and sold, like IP rights or carbon credits or conceptual art. It might also become a tyrant's perfect tool of domination, for which there were some predictions at the time.[42]

§

The story of the Facebook Like button ends where it all began. In 1941, Vannevar Bush, in his famous article on the "Memex," commented: "Such machines will have enormous appetites. One of them will take instructions and data from a roomful of girls armed with simple keyboard punches, and will deliver sheets of computed results every few minutes. There will always be plenty of things to compute in the detailed affairs of millions of people doing complicated things."[43]

But the Facebook Like button is not just the culmination of one story, it is the beginning of another one. What it unleashed has become clearer with the passing years; it is not just a clever trick for amassing data, useful for targeting ads, but an entirely new kind of public deliberation: the algorithm. It is in response to this global experiment – launched by the internet we could have had to tear down the line between public and private words – that we might even catch a glimpse of a new kind of political subjectivity.

Of course, there is no algorithm. Like the priests who translate the ravings of Pythia, algorithms are also made of people. The idea that algorithms are just software or mechanical processes obscures this reality, of course. The algorithm is a mangle of people and processes: a corporation (which itself is a person under US law) creates software rules for updating, sorting, analyzing constantly shifting data generated by billions of humans interacting and speaking and clicking and engaging, sometimes even engaging with other algorithms and bots and automated processes. It is done in the hopes of being together with other people; it is an attempt to get the attention of someone or some set of people or "everyone" – to be finally heard instead of just speaking into the void.

The animation of deliberation achieved by the Facebook Like button, and many other technologies of the internet we do have, is what sustains the algorithm: large-scale data sets, differentially accessible to platform companies and users, tuned in various opaque ways ("black-boxed") in order to achieve a variety of goals (advertising, political manipulation, conversion to a worldview). At their most elemental level, algorithms direct revenue, usually related to advertising. But people do much more than just submit: they speak, and speak differently. The process of deliberation is changed from an exchange of ideas, arguments, or persuasive techniques, into a resonant transmission of affects. Because the calculation of all that liking and sharing happens so much faster than debating, the regime of veridiction shifts from a world of statements that are true/false to a world of feels and vibes, where what matters is whether a message is loud or faint, visible or invisible – *virality*. From the world of slow reason and reasons, to the world in the moment, mindful, nudged. But it is not simply that we have moved from the world of rational deliberation to the world of behavioral economics, as clear as that transition might be. The real change is in the political subjectivity that actively seeks to *win* this new game: to win at the algorithm.

In a series of articles that appeared long after the Like button clicked its way to our souls, new media scholars have outlined elements of the way this regime of veridiction works. Drawing on earlier work from Taina Bucher, Kelley Cotter, for instance, helpfully dubbed this process "playing the visibility game" and showed how Instagram influencers think through the limits and possibilities of the algorithm from the outside. Malte Ziewtiz too has focused on the Search Engine Optimization (SEO) system for gaming advertising changes the speech and action of people in a business context.[44]

The "visibility game" is more than a question of fame and celebrity; it is much more than mere surveillance. It is an expression of a different kind of political subjectivity which submits to the algorithm, which also means: submits to the animation of deliberation. Winning the game is about tuning into a perfect resonance. The visibility game is an interpretive practice of navigating vibes, of tuning resonance, based on partial understanding of deliberately opaque algorithms, a limited form of quantitative information about engagement and views, and a lot of "research" about what messages go viral and which don't.

Most critical scholarship argues that such a win is only a reflection of the economic tuning of the algorithm by the platforms themselves: it is an "outrage machine" or an addiction engine that is designed to extract the maximum amount of attention from the maximum number of people. There is evidence, as if evidence were even needed, that this is what platforms try to do with their algorithms; though this gives them too much credit for being able to control what they have created. Most firms are also subject to the demand to care for their data and the way their users engage with it in some way.[45] If it were true that we are nothing more than passive scrollers, unable to make choices or think – doped to the gills and unable even to close our nictitating eyes – this critique would be sufficient. But that is not what everyone does: many, many people play the game, and by playing the game, they perform the algorithm.

The animation of deliberation emerged from the internet we could have had: first, out of the desire to let everyone speak; but second, out of the equally strong desire to use the power of computation and networks to navigate the scale and effects of that decision. It was an experiment that took the privately held interests of "opinion," accessible at scale only through the operations of polling and tallying, and turned it into something living: the beating heart and the sweaty palms of a debate without rules, evidence, or judges. A debate won by vibes, by virality. Which is also to say, a new way of "finding the right words at the right moment" but possibly now "making the right meaning in the right moment." It is a kind of deliberation that subjects the rationality of debate and evidence to the quantitative logic of voting at what is, for humans anyway, an infinite speed. The animation of deliberation happens across an infinitely shifting configuration of collectives: not just different "platforms" but the entangled algorithmic processing across platforms, enabled by the innovations of semantic webs, APIs, social graphs, and more.

The animation of deliberation was an experiment to combine the qualitative, embodied, hermeneutics of speech with the quantitative, computable, logic of votes. Compiled together, they produced a new kind of qualitative problem: the problem of interpreting virality. How will we come to know the difference between good and bad virality? How will we develop this new kind of judgment?

§

The animation of deliberation separates speech and action from togetherness. One version of the internet we could have had would have made it into a digital library, putting "all the world's information" at our fingertips; collecting speech and writing in an archive that would structure the possibility of debate and decision. But another version of the internet we could have had put us all in the space of appearance together, in a bid to let everyone's voice finally be heard, together, and tried to sort out the structure of these relations: who would hear what, who could be trusted, what would it mean to mean?

The desire for scalability brings people – lots and lots of people – together into a platform. The algorithm then slices up that togetherness: people occupy an uncertain but infinitely unfolding series of collectives (some we used to call public and some we used to call private). These collectives are inaccessible to human perception or understanding: they are computed instantly and automatically, with a complex arrangement of fast-moving and changing data. One moment these million people, the next a different million, whose words and images (reflecting, we say, interests, hopes, and desires) are jumbled by the algorithm and redistributed to the next collective. The interests (*inter-esse*, in Arendt's theory) – that which exists between us, and not as an opinion or a truth in an individual's mind – form the basic locus of power between people appearing together. These interests are sliced up and redistributed by the action of Liking: transformed from the ordinary human speech of two bodies in a physical space into a resonant wave washing over the eyes and ears and heartbeat and blood pressure of the scroller. Love and hate, violence and peace, good ideas and bad, ideas and critiques, call and response: these are signs of a Discourse; signs that we are in a debate amid abundant talk of polarization which, by imitation, creates more (talk of) polarization; signs that we might be heard, but not when we want to, and too often when we don't. And never together.

The algorithm combines the qualitative criticality of public speech into the quantitative logic of the voting majority. Voting is well understood: the isolated experience of expressing one's interests in the poll or the voting booth. It is comfortable as a kind of modern political subjectivity: I vote, my vote counts, is counted (I hope), and then the results are displayed, showing me what my fellow citizens believe. Critical deliberation is different, but equally comfortable: I speak my mind, I give arguments, I respond phatically or emphatically.

Combine the two and something different occurs: I Like, and what comes back to me is neither argument nor debate, but the summation of millions of minds having spoken, represented not just as a number but as the next video in the scroll, which I interpret, which I feel. My only recourse is to play the visibility game: to make another video, speak again, to document my reaction and react to that, but this time with more feeling than what I felt, edging closer to what the algorithm wants; which is to say, closer to "we."

Unlike the human logic of the public space of appearance, the human and machine are combined into an inscrutable oracle; the algorithm rewards the hero who conquers this oracle. The oracle no longer foretells the fate of the hero: the hero controls the oracle, and, thereby, fate – but only for an all-too-brief moment. The game at stake is not just one of "influencers" and celebrity: the very structure and locus of the space of appearance has been transformed. To be heard, therefore, it is necessary to represent oneself, critically, creatively, collectively, or individually in the way most likely to activate the algorithm: an algorithm manifestly designed to make visible only that which is most Liked by the largest number of people. Everything else fades away as unimportant, qualitative, intuitive, affective, personal. Only the viral convinces.

For the political subjectivity we have today, the algorithm must be attended to and observed and interpreted and constantly fed in order for us to be heard. This goes for everyone: politicians and actors as much as friends and family; dictators and bureaucrats as well as cats and bots. One no longer simply expresses an opinion; or, more precisely, expressing an opinion is no guarantee that your opinion will be heard. One has to craft, revise, refine, and precisely time one's opinion so that it is heard. Such opinions are forged in a sea of Likes and views, whether or not one engages in actively Liking or even notices the tally. This process determines what circulates and what is imitated; it determines the resonant field all around us, in ways similar to but also much different than how mass media and propaganda did before the internet we could have had came to the rescue and then let slip the rope.

Contemporary speech must be tuned to be heard, which means to be *felt*, not tuned to be understood. This new political subjectivity supplements, but does not fully replace, an older one: that autonomous, sometimes rational actor, with interests and beliefs and opinions, who votes and argues and debates and engages in politics. Arguments and evidence still matter but they are now in a new relation with the resonance of virality. For the time being, the political subject of rationality and qualitative deliberation is homeless.[46]

It doesn't much matter whether the oracle is owned by Google or by Apple or by Meta or by . . . Oracle; by scaling up speech and action, across all platforms, we have remade political subjectivity around the need to be engaged and to be heard at all costs, regardless really of what it was we had to say. Because *for the time being* it is clear it literally pays to be heard. For the older political subject, the danger was that she who could pay the most to speak would be the winner; for the new political subject, the one who is heard is the one who gets paid. The consequences of this change for liberal political theory really could not be more significant, and yet too many liberal political theorists continue to insist that nothing of any real importance has taken place.[47]

§

Tearing down the line between public and private speech broke deliberation in ways that could not have been predicted.

It was well meaning, and it continues to be well meaning. The public sphere we could have had was subject to now classic critiques centered around issues of equity, access, and justice. Public spheres, said these critiques, were structured to allow only certain kinds of speech (white, male) and excluded others.[48] The solution to a broken public sphere, it would seem, is to increase inclusion and limit exclusion, to remove bias. But it should be clear that the idea that some voices are heard and some are not, while demonstrably true, doesn't resolve the problem that the algorithm creates. These critiques and demands for inclusion, which are sophisticated, work from the position that the pre-eminent political subject is the one who speaks from a standpoint or situation, an autonomous subject who has (diverse) experiences that need to be communicated and to be heard. But this does not work if the political subject is one who speaks in order to generate an economic dividend, or because what one says, or who one is, is irrelevant, or merely a resource to be mined for economic value.

Instead of demanding to be heard because one has an experience that deserves to be recognized on its own merits, being heard has become its own goal and its own challenge. Instead of demanding to be heard because one bears witness to the injustices of powerful organizations, being heard is a callow appeal to those same organizations to shine brightly on your speech, regardless of who it condemns or what witness it bears. To be heard, one must second-guess the algorithm – which is made of people – playing a kind of lottery that will allow one to speak, and by speaking, win. The outcome is that people now cultivate rich and varied ways of being heard – they parry with the algorithm in a dance of visibility and faintness, a fight, a constant supplication, not in order to express an opinion, a desire, or a claim, but only to be the one who will be heard above all the others also speaking. Maybe it is because when one is heard, now one has the floor, and now one can say whatever one truly wants. But what one says no longer seems important because simply being heard is the reward, in a literally vital sense; so one must keep saying that to live. What more is there to say when it is the saying that keeps you alive? What would you say now, on the internet you could have had, if you could have had it?

Else Hackers

"Give the public free access to the memory and data banks."

Lyotard was not alone in wanting to give the public free access to the memory and data banks. It resonated with what others would call transparency, the "right to know" or "freedom of information." At issue is not simply free speech, the question of whether people are allowed to speak or not, or the content of that speech, but the information that is essential to governing, and the means by which that information is debated, argued, deliberated. Information and knowledge were always an issue of free speech and assembly, with an august constitutional heritage; but, as Michael Schudson asks, in the mid twentieth century: "how did information disclosure or transparency come to be identified with self-government, 'being informed' with good citizenship and openness with both public and private virtue?"[1]

The 1960s and 1970s saw a remarkable efflorescence of reforms, especially in the American context – often not well known – that supported this new set of ideals around being informed. Schudson's list includes the Freedom of Information Act (FOIA) of 1966 (and state versions of the same) which is now a key, established, and expansive bureaucratic prosthesis in every federal and state agency allowing anyone to request, though rarely receive, information about the operation of a federal or state agency. It included a range of consumer protection laws, most charismatically associated with Ralph Nader's crusades, and including per-unit pricing reforms, automotive pricing disclosure laws, and nutrition labeling. It included the Legislative Reorganization Act of 1970 which established that all votes in the US Congress would be recorded and made publicly available. And it included the National Environmental Protection Act, which established a novel attempt to balance human and non-human rights, grow the participation of the citizenry in executive rule-making, and also, in the form of the Environmental Impact Review, to make transparent and open detailed information about new developments relating to the environment. In the same period, journalism changed quite dramatically, becoming more aggressive, less deferential to elected officials, more investigative and adversarial. The Watergate scandal and the Pentagon Papers are the most storied emblems of leaking and whistleblowing from this period.

There is no single name for what Lyotard and Schudson identified at this moment: openness, transparency, privacy, anonymity, visibility, surveillance are all tied up in knots here. Censorship, copyright, access to information, leaking, whistleblowing, anonymous sources, but also the media and technology by which information circulates were part of the problem: what was once a print culture of newspapers and gossip had long since become a mass media of signals, waves, and networks. As Ithiel De Sola Pool framed it in a book from the same era: "The new communication technologies have not inherited all the legal immunities that were won for the old . . . and so as speech increasingly flows over those electronic media, the five century growth an unabridged right of citizens to speak without controls may be endangered."[2]

At its core the ideas of the right to know or the "free access to the memory and data banks" support a collective unafraid of the State; unafraid to try out new ideas, contest rules and structures of governance, speak freely, debate, dispute, live differently – all through mechanisms other than violence, or terror.

§

If there is any group of actors who might clearly heed the clarion call to "give the public free access to the memory and data banks," it would be hackers.[3]

Hacking and hackers define the internet we could have had, perhaps more than any other idea or set of actions. And yet hacking has a kind of fugitivity, or insurgency, that transcends the internet we could have had, and should not be reduced to it. Hacking was a necessary, but not a sufficient, motor of the internet we could have had: it involves both a positive imagination of a possible world and the work of bringing it into existence, as well as exploring the proper and improper uses of technologies. Hackers, by definition, choose to occupy a space outside any social contract, but most do it as a service, as it were. To stand outside the contract is to provisionally ask about how the rules are structured, how easy they are to break, and how good we are at making them work. Hackers are criminals only when they lack the strength to occupy a space outside the contract. There is a difference between being outside the law and breaking it; there is even a difference between contesting the law and ignoring it. But it is risky business. Like the trickster, the hacker is by definition outside all relations other than those with other hackers (ergo, one classic definition: you are a hacker only when another hacker calls you a hacker). Which is also why a doctor, a lawyer, or a politician cannot be a hacker.[4]

To a first approximation, "opening up the memory and data banks" is a pretty good definition of hacking. It doesn't always include "give the people free access," but that is one of its variant forms. Sometimes hacking asserts "you can make art and beauty with a computer," echoing the arts and crafts style of making in the free software movement. More often, it is an attempt to counteract the activities of those entities that restrict this freedom of the new, the open, and, most precisely, the "common" to all.

Hacking is a variant of speech and action. When Lyotard says "to speak is to fight," he raises the very question of where the line is between the two. Speaking between humans, in its ephemerality, is opposed to terror and violence, and yet it can still be fighting. But what about writing, what about programming, what about controlling the infrastructures by which speaking happens? These are the hacker's questions.

§

Lyotard did in fact give an indication of what it would mean to "give the public free access to the memory and data banks": in the footnote to that phrase, he lists a range of radical and autonomous movements in 1970s media that exemplify alternative "popular networks of multimedia communication": "amateur radios (especially their role in Quebec during the FLQ affair of October 1970 and that of the 'Front commun' in May 1972), community radios in the United States and Canada; the impact of computers on editorial work in the press; pirate radios (before their development in Italy); administrative files, the IBM monopoly, computer sabotage."[5]

Lyotard's actors are arguably hackers and pirates, or their precursors. Not all hackers hack for political purposes; some are simply criminals or fraudsters trying to exploit weaknesses in a system; some forms of hacking (these days sometimes called "ethical hacking") have long had as their goal the discovery of flaws in order to shame the makers of technology into fixing them. But it turns out that "Give the people free access" is a pretty good definition of the goals of one of the most famous groups of hackers on the internet we could have had: Anonymous.

Anonymous were a loosely associated band of hackers active from around 2008 to 2014, who started as troublemakers "doing it for the lulz" on the internet we could have had. But they soon became something more: engaging in more overtly political causes. As Gabriella Coleman explains, Anonymous created a style of "hack and leak practices" that could transform hacking from a playful form of disturbance to an activist commitment to a cause.[6]

As whistleblowers, Anonymous saw a new kind of action in access to the memory and data banks: the power of revelation, and with it of shame and accountability. As with the Pentagon Papers or a FOIA request, hacking into the memory and data banks was a direct route to achieving the same basic goals: to reveal information so that the game is fairer. The power of information was in its ability to force action: it was not in the production of knowledge or the performative effects of language that this information consisted, but in its ability to reveal what was hidden, demonstrate the truth solely by being made visible, because, for better or worse, it relies on a theory of truth as self-evident. This is also why it is so dangerously entangled with conspiracy theorizing.[7]

These "weapons of the geek," as Coleman calls them, encompass the entire history of openness and security in the internet we could have had.[8] On the one hand, they are revived and familiar forms of leaking and whistleblowing associated with journalism.[9] But on the other, they were also taken up by hackers – especially those who concerned themselves with security research on the internet we could have had – as a technique for shaming companies who produced insecure, buggy, privacy-intrusive software.[10] The orientation toward power involved using these weapons to reveal old-fashioned forms of control, corruption, and malfeasance. They are deeply aligned with the goals Lyotard laid out: no longer games of zero-sum minimax equilibrium (which is to say, games fully enforced by the lack of knowledge about what an actor is doing) but games opened up to the horizon of language itself because all the information can be made visible to anyone who wants to take a look. Hacking is, famously, about taking a look: at the insides of a computer, a file, a database.

Anonymous were anonymous for a reason. Anonymity enables a particular kind of free speech: speech disconnected from the identity and history of a person, and disconnected from a determinate context. It enables a radical focus on content alone and therefore a capacity to experiment with being completely open. The classical definition of anonymity, in its noble or virtuous mode, is that it allows things to be said which cannot be said in polite company, or for fear of reprisal, or because a censorship regime exists to shut down speech. In the best of cases, speaking anonymously harms those in power by saying what cannot be said otherwise; by revealing facts which, by their very self-evidence, bring shame or ignominy on the party they are directed to. In an older era, anonymous speech risked saying publicly what was not being said, or only said quietly. The age of "the right to know" brought changes to this function of anonymity. Now the "anonymous source" revealed what had been said, or written, but was deliberately withheld from appearing in public.

Electronic media created new twists on anonymity though. Whereas, in a previous era of publishing, anonymity had to be achieved through publication, because of the very difficulty of being anonymous in a face-to-face setting, it was still a relatively staid activity, policed in part by the gatekeepers of the media, who protected these anonymous sources. Telephones, radio broadcast, and television broadcast added to these vectors of revelation, insofar as they could be seized and used to this purpose. Leaking and whistleblowing in the twentieth century depended on anonymous sources who were accessible by phone or captured by wiretap or surreptitious recording.

The internet we could have had, including all the various draft versions of the internet, made "face to face" (interactive) textual anonymity possible in new ways.

Coleman's work documents how this particular kind of activity really took off. By observing the evolution of the imageboard 4chan, she documents the political conversion of its anonymous posters into Anonymous, and how Anonymous for a brief period (2008–14) aligned themselves with forms of socially progressive protest in a very turbulent moment. They stood against Scientology, in the shadows of Iran's green revolution, in the Arab Spring, the Occupy movement, and in coordination with Julian Assange's wikileaks project.[11]

The centrality of anonymity to the internet we could have had emerged from the get-go: by the mid-1990s, there were controversies around anonymous remailers and the laws that would govern such actions. By 2003, imageboards like 4chan were a renewed space of exploration of this controversial power: "Although there is an option to fill in a name, the community preferred the shared authorship and shunning of individual reputation."[12]

Anonymity always brings an ignoble mode with it as well: the ability to speak unaccountably – to lie, to manipulate, to be racist or misogynist without consequence; to speak from the id, not from the ego. Speaking the truth, in this context, means to force listeners to focus on the content of speech, rather than the speaker, even if that speech – regardless of who speaks it, because they are anonymous – harms others. There is, in this mode, always a concern with the fact that not all speaking is fighting; not all speaking is a way of arguing or making moves in a game. Sometimes to speak is to wound, something Lyotard tried to theorize as a form of terror: follow these rules of speech or you will not be allowed to speak. In the context of unaccountable anonymous free speech, the relevant rule is: nothing can be silenced, no matter how much it hurts another person to say it. Only anonymity protects such a rule because it neither requires nor constitutes a social bond. Where relations emerge, they govern the harm that speech can do. Anonymous, the collective, experienced this emergence and one result of this experience was that they shifted from a radical enjoyment of transgression to a political program of revelation.

Anonymous created something new. Not only was their game about giving the people free access to the memory banks, it was also action; as Arendt put it, they were "finding the right words at the right moment." They played a game of revelation – one that often also included denying speech to others through DDOS attacks – whose goals were to make sure that other kinds of statements (lies, denials, cover-ups) could no longer be made. Not only are openness and transparency at stake, so too are anonymity and privacy. The principled among Anonymous practiced the famous tagline from Wikileaks: transparency for the powerful, privacy for the weak. They sought to use their actions to both open up the secrets of the powerful and, at the same time, insist on the right to privacy, a curious combination mounted in exploration of the power of revelation.[13]

Anonymous were, for a while, heroes of an unusual sort. The kind of action that Anonymous engaged in came with its own form of glory.[14] This is a paradox insofar as it seems impossible that it could exist without naming the heroes at the heart of this action. Anonymous inverted this: an anti-celebrity, anti-fame-seeking ethic which nonetheless produced massive amounts of attention and focus on their actions. They played the visibility game first, and they played it very well.

Anonymous formed during the emergence of the Like economy and the new political subjectivity that was concerned less with the content of speech, in the form of opinions or statements, and much more with the problem of being heard. They played a kind of two-sided game: revelation on the one hand (games of denotation and prescription) and *virality* on the other (games of visibility).

Anonymous were not alone though. As Coleman has shown, the same imageboard culture that gave rise to Anonymous was also able to give birth to the alt-right troll armies of racist and misogynist users that Trump and MAGA operatives were able to cajole and entice into a far-right political agenda across a range of causes, from Neo-Nazis to mens' rights. They consolidated a conspiracy-threaded narrative of Social Justice Warriors (SJW) and Mainstream Media (MSM) who threatened America and its prosperity with the techniques of getting attention.[15]

According to Coleman, the latter did not evolve from the former: it is not the case that a liberatory technological milieu gave birth to its opposite, nor that somehow far left and far right activists are somehow creatures of the same bouillon made into two different soups, to use Coleman's gustatory metaphor. Although she doesn't put it this way, Anonymous represented the end of a certain political subjectivity we could have had, and the emergence of another that we now do have.

§

Anonymous were a collective, strangely. Maybe the very last collective of the internet we could have had, in the sense that they actually operated closely together, if hyper-secretly, out of view, and using the tools of privacy and anonymity that had emerged with the internet we could have had (Tor, Signal, IRC, Tails, and a bevy of other – still useful and ongoing – technical tools for self-protection from terror). They maintained this collectivity and attempted to use it for speech and action, during the emergence of a very different kind of political subjectivity: virality.

Anonymity is often understood as an individual right. Whether it is the anonymous participant in a market, the private individual, or the person who speaks truth to power. But speech never happens in isolation. Anonymous sources in the context of a leak are always more than individuals.

The duo journalist–leaker or journalist–whistleblower uses anonymity to protect but establishes credibility because the journalist vouches for the person being protected. The information never just stands on its own, even though the identity of the person speaking is unknown. Rather, it works its magic by circulating through at least two members of this collective of anonymity.

On the internet we could have had, anonymity was an experiment with relations. Just as with experiments in figuring out markets or societies or communities, anonymity was a way of diagnosing a problem with existing relations, and imagining into existence an alternative: anonymous remailers, encryption, cypherpunk manifestos, imageboards, and IRC. What was created by the internet we could have had was not a new form of anonymity, but the possibility of a peculiar anonymous collective, anonymous to each other as well as an audience. It is a seeming paradox: how could a bunch of people, who don't know each other, form a collective?

An anonymous collective – including both groups like Anonymous and other troll armies or vigilante formations – is in theory the opposite of a face-to-face community grounded in history and context, shared experience, and especially intimacy, to say nothing of communities grounded in identity, race, ethnicity, or land. Many such collectives get their momentum in a space of pure speech and debate, or from a purist sense of the value of free speech; embracing anonymity is sometimes about rejecting a sense of reputation or status because it is seen as interfering with the content of speech.

The internet we could have had provided Anonymous with a space of appearance where they could generate the relations that would form an anonymous collective. That space was not physical, and in fact many of the members may not have known each other's real identities, only a pseudonymous regularity. But it was nonetheless a space in which people appeared to each other as humans together (often with a concomitant suspicion that another pseudonym might be a chatbot, or worse: a federal agent). It was a space of speech-writing (IRC, email, social media), and not a physical forum like the agora (though occasional in-person protests, always in masks, were also common). That space of speech-writing is the substance of the relation as well: writing of software hacks, creating memes, making statements and plans for going viral on social media.

An anonymous collective can develop such relations as a result of their association, but usually not as a cause of it. Coleman's documentation of Anonymous demonstrates this clearly: by the end of Anonymous' most productive period, the members were deeply entangled with each other and with others who supported and cared for them, even if they remained not anonymous, but pseudonymous, to each other.[16] Anonymous gathered within specific media like message and image boards, in contact with each other directly; they also lived in public through elaborate forms of speech and shared narratives, memes, images, videos, and inside jokes.[17] What formed their bond at the outset was a particular relationship to speech and information: not who they are, but what they say – and, especially, their commitment to their right to say it. While Anonymous would go on to politicize their activities as a work of hacking and leaking, many others would take a different path: recruited, "red-pilled," transformed by affective attachments that resulted from the same power of anonymous collectivity.

The Anonymous of 2008–14 were the last collective on the internet we could have had, and the first explorers of the one we do. The internet we could have had was a reimagination of journalism, the public sphere, mass media, and markets. Anonymous were in league with the journalists, relying on their power to manipulate public opinion with evidence: "Anonymous loved to hate on establishment journalists. But not out of contempt; they expected better of them. Even as they engaged in illiberal direct action, their epistemology was fully liberal: they worked with establishment journalists to amplify their messages and get information out."[18]

Anonymous wanted more and more people to hear their message, to participate in their operations both by consuming leaked information and by participating in denials of service attacks and other ops. These activities were, at least for the core group, a way to use the internet we could have had as a kind of medium of progressive transformation of society and not as something separate from or better than the societies within which it was spreading. It was the last moment before the internet we now have.

Out of the same internet we could have had emerged something else. In the age of the algorithm, and of the game of visibility, the event of speech and its content are reversed. Rather than appealing to the game of knowledge and revelation, one appeals instead to the game of virality, the game of manipulating affect and meaning.

The MAGA operatives and alt-right troll armies of the Trump era aim not at revelation but at controlling the vibes: not logic and rationality in arguing the substance of an issue, but instead the conversion to a worldview through the control of the intensity and the circulation of meaning (virality).

Over the course of a decade, alt-right trolls on 4chan fitfully developed a different relationship to journalism than Anonymous, one that connected to politics in a different way, through worldviews and conspiracies. They constructed a complex conspiracy theory around Main Stream Media (MSM); they preferred instead a model of communication that treats pure speech not as revelatory, not a question of sunshine or transparency, but as a medium of resonant power, very much in the mold demonstrated by Trump as a communicator. They created new relations as *worldviews*, or worlds even, which develop out of the dynamics of anonymous relating: memeing, shared narratives and imagery, a shared sense of zealous inquiry (conspiracy) unmoored from previous collective kinds (the truths of family, community, nation). It requires a submission to the particular kind of manipulable, closed worlds that social media creates, but creates endlessly and algorithmically.

This new form of relating relies not on the game of revelation but on the powers of virality – in particular, on the ways it harnesses resonance. Susan Lepselter describes this power well in her study of people enmeshed in conspiracy:

> This is how many people make their poetic, epistemological, and social worlds: by tuning in to the half-rhyme, the near correlation, the parallel tropes that intimate a pattern. And then, in metatalk about that pattern, you spin it back out again into the world to meet up with another story, to make another rhyme. It makes sense, and remains mysterious at the same time.[19]

She could be describing the animation of deliberation by the algorithm (or a large language model, which is its successor), but she is referring simply to an increasingly common way certain people engage with each other.

There is creativity and autonomy in virality, but it is in the service of the game of visibility, of being heard, and not a game of revelation, or of rationality. It functions not simply within closed societies (as Popper might choose to say), but also in abstract ones: they hybridize the power of the generative with the terror of the closed society. They take full advantage of the ability to control meaning, not words.

On the internet we could have had, Anonymous could dwell securely in the rationality of the public sphere's deliberative powers, armed with the power of leaks and whistleblowers, to shed light on dodgy dealings. Journalists and major media outlets remained a powerful institution of accountability, a fourth estate. Most of the time institutions responded, the legal system sorted out the contradictions and inconsistencies, a sense of order returned with each new chaotic assault.

But the internet we do have displaced the power of that estate: the decimation of local news sources, the triumph of the Facebook timeline, the willing participation of journalists in the tweet and Like economy. In the internet we do have, mainstream journalism became just one site of disturbance and control among many: forums, social media, YouTube, blogs, TikTok, twitch, reddit, etc.

Right-wing MAGA armies vibe and leak not for truth but for meaning. They are not so interested in leaking information or relying on the rational power of revelation, but they are fully engaged in the affective power of manipulation made possible by the more Gnostic style of memes and stories and narratives, true or false, that bind them together into an anonymous collective kind. As Coleman points out, they are not interested in generating support for an operation or a change, they are interested in converting people to a worldview.

To do this, they are mining language's "inexhaustible reserve of utterances"; it is a kind of work that strangely supports conspiracy rather than simply eradicating it with sunlight, because the mode of veridiction of virality is different, less well understood.

Anonymous consisted mostly of people who thought that the power of affect would serve that of rationality, which formed the baseline of modern "abstract" societies; they bet on an affective resonance in the service of rationality: expose, shame, ridicule, debase, in order to correct behavior. But alt-right trolls and MAGA operatives bet on the reverse: that rationality can be harnessed to resonance, and become its own kind of power. Absurdity and contradiction permeate these spheres: the supposed power of the Mainstream Media to control thought comes at the very moment that the power of mainstream media is most threatened by the spread of virality as a new collective form.

Anonymous were the last gasp of the internet we could have had, but they were also the first to show us the one we have now. They didn't create either of them – that is not the role of the hacker – but they made both of them visible to us, if we choose to look.

III Future Exception Handling

What could the internet have been?

I will repeat the refrain: the internet we could have had failed for the same reason it could have succeeded. It was open, participatory, decentralized, flexible, modifiable, extensible, complex, emergent, spontaneous: a space of endless possibility. It failed not because powerful people controlled it, but because no one did.

Was this, as I have been suggesting, a failure of popular sovereignty? Popular sovereignty has no limits: its only function is to set limits on the authority, rule, violence, and domination of those who would govern, in whatever form. Perhaps that is too noble by half, but then again, the internet we could have had was one of the largest, most expensive infrastructural investments in human history. Setting aside the acrid smoke still emanating from the potlatch of venture capital, the internet let loose decades of government investment in every country and across different political periods; launched flotillas of corporate consultants FOMOing organization after organization into upgrading all manner of established systems; sewed whole-cloth corporate investment into new infrastructure; catechized high-minded support for universal broadband access across digital divides; and forced an educational revolution in which ordinary people learned, hand over thumb, new ways of typing, swiping, loving, hating, and two-factor authenticating. Even before the madness of the AI bubble, the investments were enourmous.

What if popular power had properly exercised itself in this moment? What could the internet have been? Could we have protected the internet we could have had from its errors – its deep, recalcitrant, hard-to-find bugs? What if our exception handling – our ability to put limits on the rule and authority of individuals, corporations, and states – had actually kept the program running?

In short, what would an internet without openness, decentralization, or participation have looked like?

If not openness . . . An internet without openness would not necessarily be closed, not necessarily be proprietary. Such were the things it intended to combat: the strong intellectual property rights regime of the US government; the secrecy and lack of transparency in corporations and governments; the inaccessibility of culture, of scientific information, data, and infrastructure.

I personally worked to combat such things. Between 1998 and 2003, I spent a great deal of time advocating for openness: I contributed to free software projects, I helped draft and revise the first versions of the Creative Commons licenses, I advocated for open access policies, I helped conceptualize what an open educational resource might look like as a computational object, I took part in attempts to define open data, and I advised a range of different organizations on how to do openness.

Much of this activity was devoted to debating limits. Limits on what other people could do with software or scholarship, limits on the ability to commercialize the same, limits that protected privacy, and limits that respected sovereignty. Such limits were vigorously contested at the same time as they were demanded. There were good reasons to remove limits: if you want to reserve for people the right to try out alternatives and propose new ways of being and doing, the resources of the past must be there to draw upon, they must be open, accessible. Arguing for a different future is best done on the basis of evidence and examples of its outcomes in previous cases. The reserve of language's utterances is inexhaustible, but to make the most of it, you have to give the public free access to the memory and data banks. Culture "always builds on the past" as the Creative Commons founders liked to repeat. These are also the principles and proposals that Lyotard laid out in *The Postmodern Condition*: the hope that a new era of "legitimation by paralogy" could counteract the terror of a system whose efficiency and capacity to incorporate all forms of resistance demanded that people step up their game, invent new moves using that memory and those data banks.

An internet without openness would not be closed, but would take seriously the idea of *new limits* instead of *no limits*. Such new limits were clear in the original GNU General Public License. It was serious about the limits it proposed: to keep a commons of high-quality software and encourage people to learn from and improve it without re-enclosing it. To do so meant erecting a certain barrier or limit intended to prevent crass exploitation of a commons by relying, counter-intuitively, on the protections of intellectual property law. But even that limit was insufficient because of the way that other aspects of the internet we could have had evolved and escaped the narrow confines of IP. Software escaped definitions of property (was it a thing, was it speech – was it a product, a service, a resource, or something else?). Reliant on an imposing, inertial edifice of law, the GPL could not extend its restrictions as the internet itself changed, and was changed by many hands.

Other cases made clear the cultural and historical potentials for new and different kinds of limits. One of my favorites from the internet we could have had was Mukurtu, a content management system designed by Kim Christen and Craig Dietrich designed initially to make an archive for Australian aboriginal peoples (the Warumungu at first) that would enable the use of "cultural protocols" on who could see and access which material based on values in the community.[1] Open source fanatics balked at the idea – you can't have openness if you set limits on who can use or see anything – but this was precisely the point. You can't have a viable openness if you are going to open it up to scalability without limits. Mukurtu is, happily, still going today, and still very principled, even if hemmed in by a pathological internet it yet depends upon.

In the end, making openness stand in for negative liberty is a mistake, both in practical terms and in terms of political theory. Even the thin limits that were successful in this period were too thin to be meaningful, or to force any real restrictions on the ability of people to undo those limits in the name of unlimited freedom. Making liberty work, technically or politically, still remains a problem.

If not decentralization ... An internet without decentralization would not be centralized, but would have been plural. It would not necessarily need to be the networks it was replacing: those ISO-standardized, Geneva-certified, corporate-controlled walled gardens that states and engineers insisted were the proper way to greet the future, back in the 1990s (recall the National Academies report *The Unpredictable Certainty*). The decentralization promised by the internet was at that time better understood as an archipelagic formation – a world with mainlands and isolated islands, connected in a chain of related but separate sovereignties – the very ethos at the center of a "network of networks" was to respect that plurality.

The case of the so-called "great firewall of China" or the Saudi Arabian lock-down of the internet, or just the existence of AOL or even Facebook on its own terms, was compatible with this idea: autonomous bounded networks which could interoperate and intercommunicate with each other within limits. It was a world where both China and Facebook could set the terms of how they ran their own internal network, so long as they abided by rules of interaction and inter-operation with those of other nations; the internet we could have had was essentially a technical variant of international law, trade, and diplomacy.

Where this idea met its downfall was with the continued dominance of an American globalization, which protected (or imposed, as you wish) human rights, and elevated individual choice to technical sanctity, sacrificing the diversity of forms of community, sovereignty, culture, and history to the march of a certain kind of liberatory progress. It was never enough to let well enough alone: the world had to be one.

Plurality as such would be different: a world of non-exclusive worlds remains a theoretical dream for many (it has been particularly strong in recent years in social theory, including that of Arturo Escobar, who wrote probably the earliest analysis of "Cyberia" from an anthropological perspective).[2] Perhaps the internet we could have had would have enabled a variant of these worlds, and perhaps in some ways it still does – the more complex and reticulate it becomes, the more it opens up small spaces for just such forms of autonomy and sovereignty. It is subordinated now to a larger sovereign whose power and possibilities we have yet to name; not simply a sovereign power called imperialism or colonialism, but one built out of the failures of popular sovereignty itself, and intolerant of pluralism as such.

On a more practical level, there is today still some hope for a world of "federated" social media.[3] This would not be the pluralism of many worlds, but at least it would fix some of the weaknesses of decentralization. Nonetheless, a federated media environment cannot succeed now without confronting these new games of visibility and scalability as they have come to exist; if it cannot provide new tools for an *animated deliberation*, then it will not set limits on existing domination in any meaningful way.

If not participation . . . An internet without participation would not be the autocratic system of gates and gatekeepers who control who speaks and who doesn't; not the top-down mass media it was bent on replacing, the world of broadcast networks, concentrated media industries, and censorship. I participated: I saw the how the power of the "blogosphere" worked early on, especially in academia. I witnessed how it, at first, seemed to unleash the ability of everyone to discuss and debate beyond the walls of the university or the scholarly venues; but then how it also came to be perceived as a form of gate-keeping itself, a world of influencers and virality before it had that name.

An internet without participation also might not be the world of zines and xeroxes, punk rock squats and pop-up raves, or the Myspaces and Tumblrs of yore, because all that still requires a mainstream to which to be alternative. True plurality in such a world would have to wrestle with the absence of such a thing, a river that is all tributaries, if such a thing is even possible.

An internet without participation might be a world of collectives, self-determined and persistent but which demand a kind of care and work on everyone's part to sustain; it would have to be a participation that comes with dividends both for the individual and for the entity participated in. For a while it seemed we would create such possibilities: SixDegrees, Friendster, MySpace, Orkut are the names of the social media before it became the social media we know. But the participation we might have had succumbed to an openness without limits and a decentralization without plurality.

§

The internet we could have had was more success than failure in one sense: it scaled up all the things. This is not a human success (the human costs are all too clear) but it has been an engineering and an economic success. It has been a success in the sense of what AI experts now call "reward hacking": a certain narrow set of goals were laid out in front of people and they pursued those goals with all the energy, ingenuity, capital, and competitive zeal (and greed too) they could bring to it, at every level and with remarkable speed and power. If there is a murderer, it is the drug of scalability: the endless acceleration of growth at all costs. It is a drug that has gone from being a benign Benzedrine to a mind-wrecking Fentanyl in less than a decade; openness, decentralization, and participation stand no chance by comparison.

Scaling up openness meant capturing standardization and removing it from democratic control. If the internet we could have had rested on a faith, it was that technical standards would settle the disputes of power by relying on their automatic, technical features. The most recent and most refined version of this faith emerged in the case of blockchain and crypto communities, speedrunning the internet we could have had from 2018 to 2022, only to be crushed at the Boss levels of regulated stablecoins and an AI super-centralization that will make the environmental damage of bitcoin mining look quaint by comparison. The most extreme versions of blockchain imagination share something with the original desires of the internet we could have had: that technology will replace authority, that math will become the sovereign, and because math is neutral, it will bring equality and justice with it. Popular sovereignty has never worked that way; settling disputes is never automatic, though it has had its august partisans (Leibniz comes to mind). Behind math is technological standardization, and behind that are standards organizations. And standards organizations are only as strong as the people who design and defend them; confronted with the power and speed of scalability, such organizations have proven to be not very strong.

Scaling up decentralization meant simply subordinating decentralization to another form of centralization. The case of Amazon stands here as a warning sign: there is always another move in the game, spontaneous order is not always salutary. Technology is necessary for the new move, but never sufficient to preserve the advantage. The result is something new: Amazon scaled up capitalism to subordinate whatever it was replacing; and now we have two problems. Are both things "capitalism" or have we handed over sovereignty to a new kind of entity, neither corporation nor state?

Finally, scaling up participation meant animating deliberation. It is animation in a double sense: an animated character is a stylized and exaggerated version of a real person or animal: big eyes, disproportionate, hyper-muscular. But also, animation is being *made to move*: it is the process of fooling the eyes by showing still images fast enough to produce the appearance of movement (which here means computation, specifically of relations). To animate deliberation is to do the same thing with political speech: distort it (while keeping its basic elements) and give it the appearance of movement. The game of visibility activates new modes of human perception and cognition, emotion and resonance. Personally, I've never learned to play it, and I may never. But millions of others have, and that is where the consequence lies. The question of whether we will be able to recognize, name, and harness these new modes of perception and cognition is far from certain. The scalability game could easily come to dominate by creating new modes of perception and cognition faster than any group of humans can adapt to them, outstripping the enthusiasm and energy of even those most committed to scalability itself; accelerationism would then see its fruit.

Lest anyone think scalability is a value only held in the corporate headquarters of Silicon Valley, that is not true. The power of scalability afflicts all sorts of well-meaning efforts, ranging from K-12 educational approaches to international development tools to "community"-run green energy projects. The underlying logic of veridiction is always "but will it scale?" whether that is a technology or an ethical idea. Call it the planetary demand: because the effects of scalability are global, the solutions must be too. Thinking with limits, with plurality, and with a kind of careful persistence of care are the only alternatives to the constant demand for scalability.

Else . . . The rise of the internet we could have had and emergence of hacking are often told together as if cut from the same cloth. But hacking in its most glorious forms is much older, a variant of or fugitive offspring of popular sovereignty. It is challenged by the death of the internet we could have had, but it will not die with it. To hack, in the best sense, is to occupy the space of appearance, and find the right words at the right moment, as Arendt's understanding of politics would have it, but now in a space that is no longer just that of speech and action; it is a space in which the definition of speech has changed dramatically enough to matter for the future. A politics of truth, or an ethics of justice remain central; but they are now supplemented by a pragmatics of virality and an economics of scalability. A simple politics of revelation, or the faith in speaking truth to power, confronts new limits with the death of the internet we could have had. What hackers of one generation bequeath to those of another will not be tools and techniques but the spirit of popular power.

The Brains We Could Have Had?

What would it mean, today, with renewed effort, to do as Hannah Arendt counseled and *think what we are doing*? We could do worse than to specify what we mean by thinking. Above all, thinking is something that happens together with others. We have been so bamboozled by the reduction of thought to intelligence that we have forgotten the power we have. Intelligence is real (Arendt also thought so) and Artificial Intelligence may well be a species of it.[1] But it is not thought, which happens essentially in moments when humans are together with each other.

Certain forms of thought are possible in solitude. The activity of the scholar benefits from a bit of solitude, but that bit is always in response to other moments of thought that happen together with others. The AI models we increasingly live with today cannot be together with anyone, even with each other. AI models cannot generate new statements; they cannot engage in paralogy. AI is the ultimate tool of generating the same statement over and over again in ways that appear new, but cannot shift levels or move between language games. Only a human, aware of the fact that AI is in fact confined to the software and hardware and networks and internets, only that human can find "the right words at the right moment" in a new space of appearance.

There is a story which was once lost to history: that of the ENIAC 6.[2] The women involved in the project to create the ENIAC during World War II were highly trained in mathematics and engineering, as were the men – but the women were given the job of software programming, before that practice had a name, and they recognized early on that the work was detailed, finicky, and frustrating, but also rewarding and challenging.[3] All this work served a specific, almost instantaneous moment when the program would be run like an extraordinary, mesmerizing theatrical play, condensed into seconds (well, maybe minutes, but fast for the 1940s). E for Electronic (when electrons would flow through its circuits); N for Numerical (when vacuum tubes representing numbers and mathematical operators would flicker in the huge array); I for Integrator (when the calculations set up by the women would come together to solve a problem faster than any human brain could); A for And, because you need a clever acronym; and C for Computer (the word that gives the lie to the whole thing: the machine is now called the computer, not the human women who made it into a work of art). ENIAC. Turn the electricity on, integrate the numbers and operators, compute the result, as if by magic. The women knew this magic well because they created it.

The *action* they created was captured in a single moment, a space of appearance perhaps, although also a disappearance, and a profound moment of misrecognition that has structured the story of what computers are said to be ever since. In February 1944, watched by generals, business leaders, engineers, and politicians, the ENIAC had its first demonstration. The women who made it saw the magic they produced; the men who watched saw something else. They said: "It thinks." And they were amazed by it, not them.

Nearly 100 years later, the very same show repeated itself: the story of GPT. Except this time it is not just women who will be erased from history, it is us, it is the internet we could have had. For the magic of what we currently call AI is built completely out of decades of careful, detailed, frustrating, but often rewarding work coordinated by people who wanted the internet we could have had: editors at Wikipedia, makers of free software, scanners of books and compressors of MP3s; the human labor of answering questions, taking photos, making music, all crushed into the matrix-multiplied models of AI in order to make the magical happen. All this work served a specific, almost instantaneous, moment when the model would generate a response to a human prompt, like a literally mesmerizing and hypnotic conversational partner. G for Generative because it was a model for generating text, rather than simply matching patterns and making predictions. P for Pre-trained because these models are trained, often without so-called supervision, on massive amounts of text and images, taken both with and without permission. T for transformer, the name of the "self-attention mechanism" that makes contemporary AI models so impressive.

The *action* of the GPT models was captured in a single moment, which seemed to many to be as consequential as that of ENIAC, if not yet as well and fully understood. In November 2022, watched by business leaders, generals, engineers, and politicians, GPT had its first demonstration. The internet we could have had witnessed the magic it had produced; the men who watched saw something else. They said: "It thinks." And they were amazed by it, not us.

The internet we could have had could have been a brain. The goal was clear from the start, grounded in Vannevar Bush's Memex and his article "As We May Think"; a resounding series of amplifications followed: Douglas Engelbart's "augmented reason," J. C. R Licklider's "man–computer symbiosis," Marshall McLuhan's global village and extensions of man, Alan Kay, Stewart Brand, Ted Nelson's Xanadu, the World Wide Web, Wikipedia, etc. The "global brain" was also Lyotard's "memory and data banks"; it was the NSF's digital library initiatives, and maybe also the NSA's; it was Brin and Page's page-rank algorithm; and it was at the heart of free and open everything: source code, software, data, science, hardware. The internet we could have had pitched camp on the moral high ground (an all-male panel) that was the tip of the iceberg of human scientific and scholarly endeavor.

It was more than a brain though – more than that Giant Brain that people saw in the ENIAC – because it was connected. It was an ongoing experiment with relations of all kinds, a model we put our relations inside in order to experiment with, from the very first email through the explosion of Usenet groups, bulletin boards, gophers, crawlers, webs, MUDs, and MOOs through to the Mordor of social media. Underneath it all, a military and intelligence kingdom in formation, in the secrecy of the NSA, in the hacker worlds of cracks and keys and exploits and wargames. To say, today, that Big Tech mines and extracts our social relations misses the fact that social relations have been hybridized with the media of computing – they did not exist as a resource waiting to be taken, they came into being through the explorations of the internet we could have had, of thinking together. That the internet was a next stage in human knowledge – from writing to codex to book to phonograph to type to electronic media to the computer – was an iteration on a very old set of desires and objectives.

The internet we could have had brought something new to the dream of Artificial Intelligence. Thinking is a *network of connections*, like citations in a book, or references in a work of art, or the cumulative cheerleader pyramid of giants standing on the shoulders of giants. It was a story of accumulation, connection, and complexity. Starting in the early twentieth century, a range of models of the brain figured thought as a network; the most charismatic was McCulloch and Pitts' logical calculus of the nerve net, taken up directly in the description of the von Neumann architecture at the heart of the general-purpose computer. But it also included the neural net, the perceptron, and the idea of serial and parallel processes.

The internet we could have had traded on these ideas by externalizing the image of the brain into a network of computers and memory and data banks. The internet was neither simply the source of knowledge nor the archival sink for it. It was a tool for exploring those connections: augmenting them or becoming symbiotic with them, hybridizing knowledge with relating in new ways. The internet we could have had, therefore, was an enthusiastic experiment in the exploration of knowledge in the "information society" and beyond. Not only did it afford this view of human knowledge, but it demanded that people work to bring it into existence by participating. The nature of network structure was already a concern of academics, one that was lit on fire with the realities of network structure, degrees of relation, weak links and strong, page rank, and relational database.

The internet we could have had worked hard to fill this brain with possibilities: mailing lists, databases, wikis, blogs, bulletin boards, gnutellas and napsters, microblogs and "social software," tumblrs and friendsters and myspaces, and so on. What captured the imagination was the external manipulation of these networked structures, *as relations*, as speech-writing, through software, databases, interfaces, and communication channels. Page, file, folder led to link, socket, port led to API, RSS, JSON and on. Animated knowledge became more than the sum of its parts: it was the wisdom of crowds, organized without organizations, a fever pitch of networked knowledge in a global brain.

Even better, it was all going to be open, in the sense of the "open system," precisely because this global, collective brain needed to be autonomous, uncontrolled by any nation, state, corporation, or individual (otherwise, it was a recognizable SciFi dystopia). Without autonomy of the global brain, it could fall under one entity's control, limited, cramped; with it, it would be infinitely growing, modifiable, infinitely possible. And so we made it all open, extensible; we licensed it for re-use and derivative works; we built institutions on the promise of re-use and policies on the demand for access, and futures on the desire for the unknown, which would also satisfy a desire for justice.

Was it a surprise when this structure started to be captured inside a range of containers? Extensibility subtly transformed into totality. Rather than a library with volumes perpetually added, Google digitized and controlled all the books, becoming a competitor to libraries rather than an extension of them. Rather than a network perpetually extended by individuals with their PCs, Amazon offered a cloud with the internet inside it. Rather than a space for the infinite exploration of music, Spotify emerged as an interface to all past musical memories and future ones too: wrapped by your data. Every such transformation was attended by the most intense legal and cultural contestation: property rights and licensing, monopoly and anti-trust, cross-border commerce, data sovereignty, privacy rights. The cloud and the data set and the data center ate the network and the link and the wiki. Moore's law, doubling every 18 months, was no friend to the extensible network, but it was court counselor to the consolidating platforms.

All through this period Artificial Intelligence continued with its empty promises, through summer and winter. A now standard story has it that "symbolic" AI, begun in the 1950s and 1960s, which rested on abundant research and creativity in logic and symbol processing by computers, dominated the second son, the other AI of learning, neural nets, and "machine learning," tilling the thankless but fertile fields of statistics and probability. The former is glorified in the most reductionist terms possible, capable of ultimately reducing the miracle of thought to a crystalline matrix of axioms, rules of transformation, and calculations of tables of truth, to the magic of zeros and ones. The latter could only project into the future, when the nets were faster, the data was bigger, the memories and associations connectionized. The distinction is overdrawn but the formation of distinct research communities and the fights over the right way to do AI were real.

Those forlorn AI researchers in waiting paid attention to the increasing containerization of the internet we could have had. They were waiting, making small progress with their small models and small data sets, painstakingly labeled in ways that would eventually come back to haunt them. The internet we could have had was really, after all, a giant brain, the brain they needed – a resource for something else they were only allowed to call a model or an algorithm (because the scions of "real" symbolic AI refused them the name), a second brain. Thus emerged the LLMs, the foundation models, the promises of not just AI but AGI; G is for General here, not Generative. G is for God.

But these promises are empty because that second brain is simply made up of all the speech-writing relations within that previous brain: the internet we could have had with its precise, preservable links and navigable nodes and semantically labeled transformations representing human culture and thought. It was human knowledge and it could be collected: common crawls of the internet, scanned books and archives, newly labeled data sets and unlabeled blobs of text, secret skunkwork scraping projects as well as back-room decisions to plunder everything despite the hard-won legal limits of the IP industry we once fought so hard to resist.[4] Immense containers of "content" collected by the biggest platforms who had collected all the smaller platforms who had collected all the smaller platforms and so on.

This new second brain would also contain not just the books and the archives and the IP, but also the conversations. Wikipedia, reddit, flickr, Spotify, IMDB, to say nothing of the billions of interactions on social media which were not "content" but connections among content: relations. Content created in order to facilitate those connections, the sum conversational total of a certain, limited but large, semantically labeled, computable version of human language – speech-writing. And it was all produced, or digitized, in a stunningly short period of time. All of human speech and writing (supposedly, though the digital archive is both utterly partial and totally biased) was accessible now in a handful of data sets containing chatter spanning the first 25 years of the third millennium.

Back in the late 1990s and early 2000s, when I worked furiously to ensure that content was open and free, when I argued to release every last article I wrote under free access terms and licenses, I would have predicted that the first brain would give rise to the second one eventually. The global brain of the internet we could have had was going to be the ground from which newly born humans would rise, make the world better with the patrimony of all that work, all that painstaking effort that represents the human condition. I hadn't the faintest idea that this dream would come true only as a giant text adventure game backed by the world's most expensive and environmentally destructive statistical clustering algorithm, trained on stolen goods and empty promises. But neither did anyone else.

It's not the end of the world, though. The more people come to understand what the second brain is – the one we call AI for now, the LLMs and the agential whatchamacallits – the more people will realize that it is not magic, but the result of human labor and work. What's more, we will have to remember at some point soon that all this is built upon a seemingly forgotten brain: the brain of logic and logical puzzles; the brain of Turing and Church and Gödel; the one still lurking for the time being in the CPUs and GPUs of the infinite scroll megamall built on top of it. That brain was ideal, sometimes Platonically so, a perfection not empirically derived from the doings of humans, but one potentially discoverable by them. A motivating force – an essential one – still operating behind the first and second brains, long forgotten by the barkers of the AI carnival.

The zero brain, the logical one, once upon a time made some strong claims for defining the human: as rational, as logical, syllogistically enabled to rise above the animals and the machines and reason toward infinity and perfection. Such claims were delusions. Humans are not logic machines; humans are moody, humans are contradiction-propagating, vibe-feeling interfaces between a world and a self. Humans are plural, not one. Nonetheless, we have created a world filled with logic machines: they are all around us, inside and outside our bodies, networked, pulsing with electricity, registers flipping, matrices being multiplied, connectionless datagrams being forwarded and routed, decisions made and unmade, memories outsourced to them. The inside and outside of the human condition has become like a Möbius strip: we made logic machines which re-make humans which re-make logic machines . . .

Now, for a handful of mad scientists, the new frontier is to put these logic machines inside one last brain: the fleshy one inside human skulls. A brain that follows some of the strictures of logic – when it can, mostly in the morning after a cup of coffee, and only if its microbiome will comply – but a brain somehow filled with a refracted version of the second brain of connections, networks and links, human texts and ideas, past and emerging; a brain that seems to think a bit like the third one, with its transformers and weights and backpropagation. In short, a brain that is both more and less than all three other brains in this history. A brain that is more than a brain: connected to gut microbes and electric fields, dependent on circadian rhythms of light and dark. An ultra-low-energy organ that we – some of us anyway – think we are; the animated blood flow we observe in FMRI scans, and the electrical pulses we see in EEG waveforms. We look at ourselves, looking at it, and we say "It thinks," and we are amazed.

§

On the internet we could have had the title of this book is simply *The Internet We Could Have Had*. But on the internet we do have, its title is supposed to be *The Internet We Could*

Have Had, and What You Can Do to Get It Back. But here's where it ends: you can't have it back. The internet we could have had is gone forever. What's more, you don't want it back. Reanimating the dead is never a good idea. We turn to the dead for wisdom, not for action, not for plans or creativity or thought. If we insist on reanimating it, the internet we could have had will only want one thing: brains.

The internet we could have had is dead. Long live the internet we could have had.

Notes

I 404 Not Found

1 Hannah Arendt, *On Violence* (Harcourt, Brace, Jovanovich, 1970), p. 4.

2 Glenn Greenwald, *No Place to Hide: Edward Snowden, the NSA, and the U.S. Surveillance State* (Henry Holt and Company, 2014). For the actual dirt, see https://github.com/iamcryptoki/snowden-archive.

3 Ben Tarnoff, "Opinion: The Internet Is Broken. How Do We Fix It?" *New York Times*, May 28, 2022; Ben Tarnoff, *Internet for the People: The Fight for Our Digital Future* (Verso, 2022).

4 Shoshana Zuboff, *The Age of Surveillance Capitalism: The Fight for a Human Future at the New Frontier of Power* (PublicAffairs, 2019).

5 Jonathan Crary, *Scorched Earth* (Verso Books, 2022), n.p.

6 A sample, maybe a representative one: Franklin Foer, *World without Mind: The Existential Threat of Big Tech* (Penguin Press, 2017); Byung-Chul Han, *Psychopolitics: Neoliberalism and New Technologies of Power* (Verso Books, 2017); Malcolm Harris, *Palo Alto: A History of California, Capitalism, and the World* (Little, Brown and Company, 2023); Ulises Ali Mejias and Nick Couldry, *Data Grab: The New Colonialism of Big Tech and*

How to Fight Back (University of Chicago Press, 2024); Nick Srnicek, *Platform Capitalism* (Polity, 2017); Tim Wu, *The Curse of Bigness: Antitrust in the New Gilded Age* (Columbia Global Reports, 2018); Astra Taylor, *The People's Platform: Taking Back Power and Culture in the Digital Age* (Metropolitan Books, 2014).

7 Cory Doctorow, "'Enshittification' Is Coming for Absolutely Everything," *Financial Times*, February 7, 2024; Cory Doctorow, *Enshittification: Why Everything Suddenly Got Worse and What to Do About It* (Verso Books, 2025). To his immense credit, Doctorow's scatological imagination has been consistent – he also penned a widely read critique of the semantic web called "Metacrap" in 2003.

8 Cory Doctorow, *The Internet Con: How to Seize the Means of Computation* (Verso Books, 2023), p. 25.

Syntax Error

1 National Research Council, *The Unpredictable Certainty: Information Infrastructure through 2000* (The National Academies Press, 1996).

2 On the long history of popular sovereignty, I recommend Daniel Lee, *Popular Sovereignty in Early Modern Constitutional Thought* (Oxford University Press, 2016).

3 Hannah Arendt, *The Human Condition* (University of Chicago Press, 1998), p. 3.

Insufficient Memory

1 Tarnoff, *Internet for the People*; Zuboff, *The Age of Surveillance Capitalism*; Mejias and Couldry, *Data Grab*; Srnicek, *Platform Capitalism.*

2 Sarah Jeong, *The Internet of Garbage* (Vox Media, Inc., 2018); Safiya Noble, *Algorithms of Oppression: How Search Engines Reinforce Racism* (New York University Press, 2018); Cathy O'Neil, *Weapons of Math Destruction: How Big Data Increases Inequality and Threatens Democracy* (Crown, 2016).

3 The earliest: Alexander R. Galloway, *Protocol: How Control Exists after Decentralization* (MIT Press, 2006). On hubris, see Morgan G. Ames, *The Charisma Machine: The Life, Death, and Legacy of One Laptop per Child* (MIT Press, 2019). On diversity, Christina Dunbar-Hester, *Hacking Diversity: The Politics of Inclusion in Open Technology Cultures* (Princeton University Press, 2020). On gender, Joseph Reagle, "'Free as in Sexist?' Free Culture and the Gender Gap," *First Monday*, 18.1 (2013); Heather Ford and Judy Wajcman, "'Anyone Can Edit', Not Everyone Does: Wikipedia's Infrastructure and the Gender Gap," *Social Studies of Science*, 47.4 (2017), pp. 511–27.

4 Richard Barbrook and Andy Cameron, "The Californian Ideology," *Science as Culture*, 6.1 (1996), pp. 44–72; Jodi Dean, "Communicative Capitalism: Circulation and the Foreclosure of Politics," *Cultural Politics*, 1.1 (2005), pp. 51–74; Siva Vaidhyanathan, *The Googlization of Everything (And Why We Should Worry)* (University of California Press, 2012); Trebor Scholz, *Digital Labor: The Internet as Playground and Factory* (Routledge, 2013); James Brook and Iain A. Boal, *Resisting the Virtual Life: The Culture and Politics of Information* (City Lights Books, 1995). On Barbrook, see the recent reconsideration of this early and very influential essay in Nathan Schneider, "Afterlives of the California Ideology|'Polemic Becomes Canon': An Interview with Richard Barbrook on the Californian Ideology," *International Journal of Communication*, 17 (2023), p. 12.

5 Jonathan Zittrain, *The Future of the Internet: And How to Stop It* (Penguin UK, 2009).

6 *Your Computer Is on Fire*, ed. Thomas S. Mullaney and others (MIT Press, 2021).

7 Kevin Driscoll agrees: "The history of the internet could be a thrilling tale inclusive of many thousands of networks, big and small, urban and rural, commercial and voluntary. Instead, it is repeatedly reduced to the story of the singular ARPANET" – Kevin Driscoll, *The Modem World: A Prehistory of Social Media* (Yale University

Press, 2022), p. 194. Even a better history of ARPANET remains to be written.

8 First time as tragedy: Walter Isaacson, *Steve Jobs* (Simon & Schuster, 2011). Second time as farce: Walter Isaacson, *Elon Musk* (Simon & Schuster, 2023).

9 Driscoll, *The Modem World*; Joy Lisi Rankin, *A People's History of Computing in the United States* (Harvard University Press, 2018); Laine Nooney, *The Apple II Age: How the Computer Became Personal* (University of Chicago Press, 2023); Charlton D. McIlwain, *Black Software: The Internet and Racial Justice, from the AfroNet to Black Lives Matter* (Oxford University Press, 2020).

10 Such a list risks greater offense by exclusion the longer it becomes: Lisa Nakamura, *Cybertypes: Race, Ethnicity, and Identity on the Internet* (Routledge, 2013); Lisa Nakamura, "Race in/for Cyberspace: Identity Tourism and Racial Passing on the Internet," *Works and Days*, 13.1–2 (1995), pp. 181–93; Marisa Elena Duarte, *Network Sovereignty: Building the Internet across Indian Country* (University of Washington Press, 2017); *Captivating Technology: Race, Carceral Technoscience, and Liberatory Imagination in Everyday Life*, ed. Ruha Benjamin (Duke University Press, 2019); Ruha Benjamin, *Race after Technology: Abolitionist Tools for the New Jim Code* (John Wiley & Sons, 2019); Wendy Hui Kyong Chun, *Control and Freedom: Power and Paranoia in the Age of Fiber Optics* (MIT Press, 2006); Wendy Hui Kyong Chun, *Programmed Visions: Software and Memory* (MIT Press, 2011); Avery Dame-Griff, *The Two Revolutions: A History of the Transgender Internet* (New York University Press, 2023); Alondra Nelson, Thuy Linh N. Tu, and Alicia Headlam Hines, *Technicolor: Race, Technology, and Everyday Life* (New York University Press, 2001); Daniel Greene, "Landlords of the Internet: Big Data and Big Real Estate," *Social Studies of Science*, 52.6 (2022), pp. 904–27; Daniel Greene, "Discovering the Divide: Technology and Poverty in the New Economy," *International Journal of Communication*, 10.0 (2016), p. 20.

11 Nick Seaver, "Captivating Algorithms: Recommender Systems as Traps," *Journal of Material Culture*, 24.4 (2018), pp. 421–36;

Alberto Corsín Jiménez and Chloe Nahum-Claudel, "The Anthropology of Traps: Concrete Technologies and Theoretical Interfaces," *Journal of Material Culture*, 24.4 (2019), pp. 383–400.

12 Silicon Valley does have significant auto-hagiographical power – e.g. the WELL (the "Whole Earth 'Lectric Link") and its denizens: see Fred Turner, *From Counterculture to Cyberculture: Stewart Brand, the Whole Earth Network, and the Rise of Digital Utopianism* (University of Chicago Press, 2008). See also Steven Levy, *Hackers: Heroes of the Computer Revolution* (Anchor Press / Doubleday, 1984), and the WELL's autohagiography, Katie Hafner, *The Well: A Story of Love, Death, and Real Life in the Seminal Online Community* (Carroll & Graf, 2001).

13 See, for example, Luke Stark, "Here Come the 'Computer People': Anthropomorphosis, Command, and Control in Early Personal Computing," *IEEE Annals of the History of Computing*, 42.4 (2020), pp. 53–70.

14 Shoshana Zuboff, *In the Age of the Smart Machine: The Future of Work and Power* (Basic Books, 1988); Nooney, *The Apple II Age*.

15 Zuboff, *The Age of Surveillance Capitalism*, p. 9, italics in original.

16 Zuboff, *The Age of Surveillance Capitalism*, p. 505. The refrain is repeated often: "Surveillance capitalism feeds off every aspect of every human's experience"; "Surveillance capitalists quickly realized they could do anything they wanted. And they did"; "They were protected by the inherent illegibility of the automated processes that they rule, the ignorance that these processes breed, and the sense of inevitability that they foster" (9–10).

17 Zuboff, *The Age of Surveillance Capitalism*, p. 499.

18 Marion Fourcade and Kieran Joseph Healy, *The Ordinal Society* (Harvard University Press, 2024).

19 Misrecognition might be contagious: Healy was one of the most visible and creative, highly online participants in the culture of data analysis, open source, and its ecosystem (cf. Kieran Healy and Alan Schussman, "The Ecology of Open-Source Software Development" [2003; unpublished ms. available at https://kieranhealy.org/files/drafts/oss-activity.pdf]).

20 Nathaniel Tkacz, *Wikipedia and the Politics of Openness* (University of Chicago Press, 2015).
21 Karl R. Popper, *The Open Society and Its Enemies*, 2 vols. (Princeton University Press, 2020).
22 Arendt, *The Human Condition*, p. 9.

Reloading . . .

1 Galloway, *Protocol*; Vint Cerf and Robert Kahn, "A Protocol for Packet Network Intercommunication," *IEEE Transactions on Communications*, 22.5 (1974), pp. 637–48; David S. Isenberg, "The Dawn of the 'Stupid Network,'" *netWorker*, 2.1 (1998), pp. 24–31; David D. Clark, "The Contingent Internet," *Daedalus*, 145.1 (2016), pp. 9–17; Tarleton Gillespie, "Engineering a Principle: 'End-to-End' in the Design of the Internet," *Social Studies of Science*, 36.3 (2006), pp. 427–57; Lawrence Lessig, *Code: And Other Laws of Cyberspace* (Basic Books, 1999). It was very common in journalistic accounts of the internet in the 1980s and 1990s to highlight the TCP/IP protocols or the basic idea of a decentralized packet-switched network as defining evidence of its difference from other networking technologies.
2 Close in style to Dewey's definition of the emergence of the State in John Dewey, *The Public and Its Problems* (H. Holt and Company, 1927).
3 "Noosphere" came directly from a famous mystic, Teilhard de Chardin (developed with Vladimir Vernadsky) in Pierre Teilhard de Chardin, *The Human Phenomenon* (Sussex Academic Press, 1999).
4 Clark, "The Contingent Internet"; David D. Clark, *Designing an Internet* (MIT Press, 2018); Jerome H. Saltzer, David P. Reed, and David D. Clark, "End-to-End Arguments in System Design," *ACM Transactions on Computer Systems*, 2.4 (1984), pp. 277–88.
5 James Gleick, *Chaos: Making a New Science* (Viking, 1987); Ilya Prigogine and Isabelle Stengers, *Order out of Chaos: Man's New Dialogue with Nature* (Bantam Books, 1984); Ralph Abraham, *Chaos, Gaia, Eros: A Chaos Pioneer Uncovers the Three Great*

Streams of History (HarperSanFrancisco, 1994); John H. Holland, *Hidden Order: How Adaptation Builds Complexity* (Addison-Wesley, 1995); Robert M. Axelrod and Michael D. Cohen, *Harnessing Complexity: Organizational Implications of a Scientific Frontier* (Basic Books, 2000). Anthropologist Stephen Lansing provides an overview: J. Stephen Lansing, "Complex Adaptive Systems," *Annual Review of Anthropology*, 32 (2003), pp. 183–204. Helmreich captured work, at the Santa Fe Institute HQ, on complexity and artificial life: Stefan Helmreich, *Silicon Second Nature: Culturing Artificial Life in a Digital World* (University of California Press, 1998).

6 Bernardo Huberman's edited collection *The Ecology of Computation* was an early diagnosis: "A new form of computation is emerging. Propelled by advances in software design and increasing connectivity, distributed computational systems are acquiring characteristics reminiscent of social and biological organizations. These open systems, self-regulating entities in their overall behavior are quite different from conventional computers, engage in asynchronous computation of very complex tasks, while their agents spawn processes in other machines whose total specification is unknown to them. These agents also make local decisions based on both imperfect knowledge about the system, and on information for which at times is inconsistent and delayed. They thus become a community of concurrent processes which, in their interactions, strategies, and competition for resources, behave like whole ecologies" – Bernardo A. Huberman, "The Ecology of Computation," in *The Ecology of Computation*, ed. Bernardo A. Huberman (Elsevier Science Publishers, 1988), p. 1.

7 David Clark, "The Design Philosophy of the Darpa Internet Protocols," *ACM SIGCOMM Computer Communication Review*, 18.4 (1988), pp. 106–14.

8 Marjory S. Blumenthal and David D. Clark, "Rethinking the Design of the Internet: The End-to-End Arguments Vs. The Brave New World," *ACM Transactions on Internet Technology*, 1.1 (2001), pp. 70–109.

9 Michael Hauben and Ronda Hauben, *Netizens: On the History and Impact of Usenet and the Internet* (IEEE Computer Society Press, 1997); Bryan Pfaffenberger, "'A Standing Wave in the Web of Our Communications': Usenet and the Socio-Technical Construction of Cyberspace Values," in *From Usenet to CoWebs: Interacting with Social Information Spaces*, ed. Christopher Lueg and Danyel Fisher (Springer, 2003), pp. 20–43; Bryan Pfaffenberger, "'If I Want It, It's OK': Usenet and the (Outer) Limits of Free Speech," *The Information Society*, 12.4 (1996), pp. 365–86; David A. Grier and M. Campbell, "A Social History of Bitnet and Listserv, 1985–1991," *IEEE Annals of the History of Computing*, 22.2 (2000), pp. 32–41; Driscoll, *The Modem World.*

10 Joy Rankin details some of these draft versions – but narrates them as draft versions of personal computing, not quite of the internet: Rankin, *A People's History of Computing in the United States.* Her story charts experiments with networked computing in 1965–75 in which *personal computing* was figured out by citizens, not consumers, and which was displaced by the personal computer as a consumer gadget in the "transition from every student having a network to every student having a computer" (238). The story upends the popular mythology of the home-brewed personal computer of Steve Jobs – but it might also have been a displacement of the same energy to other experiments. Driscoll documents, for instance, the robust development of BBS (Bulletin Board Systems) culture as a proto-social media – more vibrant than the internet we could have had in its time: Driscoll, *The Modem World.*

11 Donna Jeanne Haraway, "A Cyborg Manifesto: Science, Technology, and Socialist-Feminism in the Late Twentieth Century," in *Simians, Cyborgs and Women: The Reinvention of Nature* (Routledge, 1985).

12 Arendt, *The Human Condition*, p. 260.

13 Marilyn Strathern, *Relations: An Anthropological Account* (Duke University Press, 2020). The choice of "relation" is deliberate here, since the resonance between being related and making relations

that Strathern describes is what people were doing when they made the internet we could have had.

14 Elinor Ostrom, *Governing the Commons: The Evolution of Institutions for Collective Action* (Cambridge University Press, 1990).

15 Manuel Castells, *The Rise of the Network Society*, Information Age, 1, 2nd ed. (Blackwell Publishers, 2000); Paul J. DiMaggio and Walter W. Powell, "The Iron Cage Revisited: Institutional Isomorphism and Collective Rationality in Organizational Fields," *American Sociological Review*, 48.2 (1983), pp. 147–60.

16 Michael Walzer, *Spheres of Justice: A Defense of Pluralism and Equality* (Basic Books, 1983).

17 John S. Dryzek, *Deliberative Democracy and Beyond: Liberals, Critics, Contestations* (Oxford University Press, 2000); James Bohman and William Rehg, *Deliberative Democracy: Essays on Reason and Politics* (MIT Press, 1997).

18 Guillermo Gómez-Peña, *Ethno-Techno: Writings on Performance, Activism and Pedagogy* (Taylor & Francis, 2013); B. Laurel, *Computers as Theatre* (Addison-Wesley, 2014).

19 Clayton M. Christensen, *The Innovator's Dilemma: When New Technologies Cause Great Firms to Fail* (Harvard Business School Press, 1997).

20 Haraway, "A Cyborg Manifesto"; Bruno Latour, *Reassembling the Social: An Introduction to Actor-Network-Theory* (Oxford University Press, 2007), p. 328; Bruno Latour, "On Recalling Ant," *Sociological Review*, 47.1 (1999), pp. 15–25. See also the previously unpublishable article Christopher M. Kelty, "Against Networks (and Still against Networks)," *Spheres: A Journal for Digital Culture*, 1.1 (2015): https://spheres-journal.org/issue/1-politics-after-networks.

21 Universal here in the Kantian sense. Cosmopolitical perhaps, but that is too ecumenical. Universal does not mean totalitarian but it does imply the forms of assent to a universal standard or a universal protocol, not unlike the categorical imperative.

22 Orit Halpern, *Beautiful Data: A History of Vision and Reason*

since 1945 (Duke University Press, 2015); Andrew Pickering, *The Cybernetic Brain: Sketches of Another Future* (University of Chicago Press, 2010). The universality of the computer appealed early on to those who saw in it a system of control, see esp. Eden Medina, *Cybernetic Revolutionaries: Technology and Politics in Allende's Chile* (MIT Press, 2011); on CyberSyn and prehistory, see David A. Mindell, *Between Human and Machine: Feedback, Control, and Computing before Cybernetics* (Johns Hopkins University Press, 2002).

23 Laurel, *Computers as Theatre*; Janet Horowitz Murray, *Hamlet on the Holodeck: The Future of Narrative in Cyberspace* (The Free Press, 1997).

24 It cannot be an accident that anthropologists were especially drawn to the idea of a virtual world: Tom Boellstorff, *Coming of Age in Second Life: An Anthropologist Explores the Virtually Human* (Princeton University Press, 2008); Tom Boellstorff, Bonnie Nardi, and Celia Pearce, *Ethnography and Virtual Worlds: A Handbook of Method* (Princeton University Press, 2012); Thomas Malaby, *Making Virtual Worlds: Linden Lab and Second Life* (Cornell University Press, 2009).

25 Sherry Turkle, *The Second Self: Computers and the Human Spirit* (Simon and Schuster, 1984); Sherry Turkle, *Life on the Screen: Identity in the Age of the Internet* (Simon & Schuster, 1995); *Communities in Cyberspace*, ed. Peter Kollock and Marc Smith (Routledge, 1998); Allucquère Rosanne Stone, *The War of Desire and Technology at the Close of the Mechanical Age* (MIT Press, 1996); Mark Dery, *Escape Velocity: Cyberculture at the End of the Century* (Grove Press, 1996).

II Have You Tried Turning It Off and On Again?

1 Arendt, *The Human Condition*, p. 199.

2 Arendt, *The Human Condition*, p. 26.

3 Arendt, *The Human Condition*, p. 52.

4 Arendt, *The Human Condition*, p. 237.

5 Jean-François Lyotard, *The Postmodern Condition: A Report on Knowledge* (University of Minnesota Press, 1984), p. 67.
6 Lyotard, *The Postmodern Condition*, p. 15.
7 Mill addressed the same processes in resisting social conformity: John Mill, *On Liberty and Other Essays* (Oxford University Press, 2008).
8 "What Is Postmodernism?" p. 81, included in Lyotard, *The Postmodern Condition*. Lyotard's appeal – very 1980s – to the "language games" of "perfect information" and the risk of "minimax equilibrium" is a gesture toward the linguistic nature of information technologies, but also Wittgenstein's language games, game theory, and the "performativity principle" governing the Cold War realities of cost–benefit analysis and the management of the enormous bureaucracies of the twentieth century. It reflected his on-again off-again debates with Habermas about communicative rationality, consensus, and dissensus. See also Derrida on this topic: Jacques Derrida, "Force of Law: The 'Mystical Foundation of Authority,'" in *Deconstruction and the Possibility of Justice* (Routledge, 2016), pp. 3–67.
9 Lyotard, *The Postmodern Condition*, p. 46.
10 Philip Pettit, *Republicanism: A Theory of Freedom and Government* (Oxford University Press, 1997); J. Matthew Hoye and Jeffrey Monaghan, "Surveillance, Freedom and the Republic," *European Journal of Political Theory*, 17.3 (2015), pp. 343–63; Christopher Kelty, "The Fog of Freedom," in *Media Technologies: Essays on Communication, Materiality, and Society*, ed. Tarleton Gillespie, Kirsten Foot, and Pablo Boczkowski (MIT Press, 2014).
11 Ruha Benjamin, *Captivating Technology: Race, Carceral Technoscience, and Liberatory Imagination in Everyday Life* (Duke University Press, 2019); Seaver, "Captivating Algorithms."

If Openness . . .

1 Sébastian Broca has written beautifully on free software as "concrete utopia": Sébastien Broca, *Utopie du logiciel libre: du bricolage*

informatique à la réinvention sociale (Éditions le Passager clandestin, 2013).

2 Christopher M. Kelty, *Two Bits: The Cultural Significance of Free Software* (Duke University Press, 2008), ch. 3.

3 Also, disputes and problems that have lasted into the present – cf. Nathan Schneider, "The Tyranny of Openness: What Happened to Peer Production?" *Feminist Media Studies*, 22.6 (2022), pp. 1411–28.

4 William Morris, *News from Nowhere* (Cambridge University Press, 1995). Eileen Boris, *Art and Labor: Ruskin, Morris, and the Craftsman Ideal in America* (Temple University Press, 1986).

5 https://en.wikipedia.org/wiki/History_of_sentence_spacing#French_and_English_spacing.

6 Aldous Huxley posed a curious critique of Morris' disdain of industrial objects: the rise in the mid twentieth century of democratic tools and tinkering. Morris' values, Huxley said, did not leave space for a demographic transition enabled by standardized goods and technologies with which men could tinker in their standardized garages. One could argue that the same kind of transition happened with the personal computer: rather than a Morris-like utopia where each crafted their own computer, instead a computer in every den, One Laptop Per Child, prosperity for all, became the mantra. Instead of one well-built computer, we found we needed several, constantly improving, disposable computers that limited our access or their modifiability in favor of convenience and ubiquity. The fact that we have the forms of addiction and waste we do today reflects back to us the chemical and industrial success that Huxley was celebrating: neither of them are making anyone healthier or happier anymore – "Liberty, Quality, Machinery," in Aldous Huxley, *Tomorrow and Tomorrow and Tomorrow, and Other Essays* (New York: Signet, 1956).

7 On the crisis of software maintenance in open source software, see: Nadia Eghbal, *Working in Public: The Making and Maintenance of Open Source Software* (Stripe Press, 2020); Schneider, "The Tyranny of Openness."

8 Kelty, "The Fog of Freedom."

9 Kelty, *Two Bits*, p. 7. See also Nathan Schneider, "Decentralization: An Incomplete Ambition," *Journal of Cultural Economy*, 12.4 (2019), pp. 265–85.

10 On the one hand, the label "peer production" maintained this hope: Michel Bauwens, Vasilis Kostakis, and Alex Pazaitis, *Peer to Peer: The Commons Manifesto* (University of Westminster Press, 2019); platform cooperativism represents a variant (https://platform.coop), see also: Trebor Scholz and Nathan Schneider, *Ours to Hack and to Own: The Rise of Platform Cooperativism, a New Vision for the Future of Work and a Fairer Internet* (OR Books, 2017). For a situated case of the long-term embrace of these ideals, see Alberto Corsin Jiménez and Adolfo Estalella, *Free Culture and the City: Hackers, Commoners, and Neighbors in Madrid, 1997–2017* (Cornell University Press, 2023).

11 Zhang and Carpano focus on the Chromium browser as an example of this and call it "logistical power" – the name is good enough but risks confusion with the logistics industry, something Amazon has clearly dominated in many ways. The "logistical" here is really more infrastructural and coordinating; it concerns decisions about where to draw the line between core and peripheral functions of a technology, as well as who gets to draw that line and how: Bolun Zhang and Davide Carpano, "Chromium as a Tool of Logistical Power: A Material Political Economy of Open-Source," *Big Data & Society*, 10.1 (2023), pp. 1–12.

12 Zhang and Carpano, "Chromium as a Tool of Logistical Power," pp. 2–3.

13 Tiziana Terranova, "Free Labor: Producing Culture for the Digital Economy," *Social Text*, 18.2 (2000), pp. 33–58; Christian Fuchs, *Digital Capitalism: Media, Communication and Society*, Vol. III (Routledge, 2021).

14 Activists like Coraline Ada Ehmke and movements like the shortlived 996.ICU movement in China raised issues of both identity and worker politics: Xiaotian Li, "The 996.Icu Movement in China: Changing Employment Relations and Labour Agency in the

Tech Industry," *Made in China Journal*, 4.2 (2019), pp. 54–9. New forms of licensing emerged on the model of free software licenses but with very different goals and controversies; see Schneider, "The Tyranny of Openness."

15 Anne Barron, "Free Software Production as Critical Social Practice," *Economy and Society*, 42.4 (2013), pp. 597–625.

16 Zhang and Carpano, "Chromium as a Tool of Logistical Power," p. 5.

17 Zhang and Carpano, "Chromium as a Tool of Logistical Power," p. 10.

18 David Gray Widder, Meredith Whittaker, and Sarah Myers West, "Why 'Open' AI Systems Are Actually Closed, and Why This Matters," *Nature*, 635.8040 (2024), pp. 827–33.

If Decentralization . . .

1 Samuel P. Huntington, *The Clash of Civilizations and the Remaking of World Order* (Simon & Schuster, 2011); Anthony Giddens, *Runaway World: How Globalization Is Reshaping Our Lives*, [2nd ed.] (Routledge, 2003); Immanuel Maurice Wallerstein, *The Modern World-System* (University of California Press, 2011); Saskia Sassen, "The Global City: Introducing a Concept," *Brown Journal of World Affairs*, 11.2 (2005), pp. 27–43; Arjun Appadurai, *Modernity at Large: Cultural Dimensions of Globalization*, Public Worlds (University of Minnesota Press, 1996).

2 Biao Xiang, *Global "Body Shopping": An Indian Labor System in the Information Technology Industry* (Princeton University Press, 2007).

3 Francis Fukuyama, *The End of History and the Last Man* (Free Press, 2006).

4 Stephen Collier, *Post-Soviet Social: Neoliberalism, Social Modernity, Biopolitics* (Princeton University Press, 2011).

5 Noam Chomsky and Robert Waterman McChesney, *Profit over People: Neoliberalism and Global Order* (Seven Stories Press, 1999).

6 Marcos, Subcomandante, *Shadows of Tender Fury: The Letters and Communiqués of Subcomandante Marcos and the Zapatista Army of National Liberation*, trans. Frank Bardacke, Leslie López, and the Watsonville (Calif.) Human Rights Committee (Monthly Review Press, 1995).
7 John Arquilla, *The Advent of Netwar* (Rand Corporation, 1996).
8 Marion Nestle, "The Selling of Olestra," *Public Health Reports*, 113.6 (1998), p. 508.
9 Schneider, "Decentralization."
10 See esp. Charles Sabel as a case: Joshua Cohen and Charles Sabel, "Directly-Deliberative Polyarchy," *European Law Journal*, 3.4 (1997), pp. 313–42.
11 Style in the sense Hacking gave it: Ian Hacking, "'Style' for Historians and Philosophers," *Studies in History and Philosophy of Science Part A*, 23.1 (1992), pp. 1–20.
12 Gleick, *Chaos*; Prigogine and Stengers, *Order out of Chaos*; Abraham, *Chaos, Gaia, Eros*; Holland, *Hidden Order*; Axelrod and Cohen, *Harnessing Complexity*.
13 Rodney A. Brooks, "Intelligence without Representation," *Artificial Intelligence*, 47.1 (1991), pp. 139–59. Brooks was one of the subjects of Errol Morris' film *Fast, Cheap and Out of Control* (1997). See also Heinz von Foerster, *Observing Systems*, 2nd ed. (Intersystems Publications, 1984); Francisco J. Varela, Evan Thompson, and Eleanor Rosch, *The Embodied Mind: Cognitive Science and Human Experience* (MIT Press, 1991).
14 National Research Council, *The Unpredictable Certainty*, p. 13.
15 Three favorite examples, of many more: Kurt Lewin and his theories of "democratic atmospheres," Malinowski's late-life reflections on freedom, and Karl Popper's *Open Society and Its Enemies* – Kurt Lewin, *Field Theory in Social Science: Selected Theoretical Papers* (Greenwood Press, 1975); Popper, *The Open Society and Its Enemies*; Bronisław Malinowski, *Freedom and Civilization* (Routledge, 2015).
16 Michael Hardt and Antonio Negri, *Empire* (Harvard University Press, 2001); Arquilla, *The Advent of Netwar*. Hardt and Negri

directly engaged open source ideals in Michael Hardt and Antonio Negri, *Multitude: War and Democracy in the Age of Empire* (The Penguin Press, 2004).

17 James W. Cortada, *IBM: The Rise and Fall and Reinvention of a Global Icon* (MIT Press, 2023); Franklin M. Fisher, John J. McGowan, and Joen E. Greenwood, *Folded, Spindled, and Mutilated: Economic Analysis and U.S. v. IBM* (MIT Press, 1983). On "1984," see Laurence Raw, *The Ridley Scott Encyclopedia* (Scarecrow Press, 2009), p. 21.

18 Paul A. David, "Clio and the Economics of QWERTY," *American Economic Review*, 75.2 (1985), pp. 332–7; S. J. Liebowitz and Stephen E. Margolis, "The Fable of the Keys," *Journal of Law & Economics*, 33.1 (1990), pp. 1–25; S. J. Liebowitz and Stephen E. Margolis, "Path Dependence, Lock-In, and History," *Journal of Law, Economics, & Organization*, 11.1 (1995), pp. 205–26; Christopher M. Kelty, "Liberty and Lock-In: The Trouble with Freedom in Anthropology," in *Freedom in Practice: Governance, Autonomy and Liberty in the Everyday*, ed. Moises Silva and Huon Wardle (Routledge, 2017), pp. 164–86. Much of this debate happened within the then vibrant discussion of a new institutional economics as a counterdiscourse to the purist definitions of the free market: Douglass C. North, *Institutions, Institutional Change and Economic Performance* (Cambridge University Press, 1990).

19 Among other key debates at the time – polycentricity: Elinor Ostrom, "Beyond Markets and States: Polycentric Governance of Complex Economic Systems," *American Economic Review*, 100.3 (2010), pp. 641–72; flexible specialization and second industrial revolution: Michael Piore and Charles Sabel, *The Second Industrial Divide* (Basic Books, 1986); post-Fordism: Ash Amin, *Post-Fordism: A Reader* (John Wiley & Sons, 2011).

20 Marc Berg and Geoffrey Bowker, "The Multiple Bodies of the Medical Record," *Sociological Quarterly*, 38.3 (1997), pp. 513–37; Stephen R. Barley, "Technology as an Occasion for Structuring: Evidence from Observations of CT Scanners and the Social Order of Radiology Departments," *Administrative Science Quarterly*, 31.1

(1986), pp. 78–108; Joseph A. November, *Biomedical Computing: Digitizing Life in the United States* (JHU Press, 2012); Joseph November, "Ask Your Doctor . . . About Computers," *IEEE Annals of the History of Computing*, 38.1 (2016), pp. 3–5.

21 Quoting John Gilmore: Howard Rheingold, *The Virtual Community, Revised Edition: Homesteading on the Electronic Frontier* (MIT Press, 2000).

22 This was a line, remarkably in hindsight, actually uttered by Joe Rogan on an episode of *News Radio* in 1995 (www.youtube.com/watch?v=0T-CreVC_6Y). It circulated widely on mailing lists as an argument about transparency and privacy on the internet.

23 Andrew L. Russell, *Open Standards and the Digital Age: History, Ideology, and Networks* (Cambridge University Press, 2014); Susanne K. Schmidt and Raymund Werle, *Coordinating Technology: Studies in the International Standardization of Telecommunications* (MIT Press, 1998).

24 Hakim Bey / Peter Lamborn Wilson is perhaps the most well known: Hakim Bey, *T.A.Z.: The Temporary Autonomous Zone, Ontological Anarchy, Poetic Terrorism* (Autonomedia, 2003). Safer for consumption were some of the counter-cultural variants – Turner identified certain geneaologies: Turner, *From Counterculture to Cyberculture*. See also Benjamin Charles Germain Lee, "The Singularity Prophets," *Current Affairs*, July 23, 2020; A. Farman, *On Not Dying: Secular Immortality in the Age of Technoscience* (University of Minnesota Press, 2020); *The Cyborg Handbook*, ed. Chris Hables Gray (Routledge, 1995).

25 Clark, *Designing an Internet*; Gillespie, "Engineering a Principle"; Isenberg, "The Dawn of the 'Stupid Network.'"

26 Wu, *The Curse of Bigness*; Lessig, *Code*.

27 See chapters 4 and 5 in Kelty, *Two Bits*.

28 See, e.g., Metcalfe's law: https://en.wikipedia.org/wiki/Metcalfe's_law. On path dependency, the classic work is W. Brian Arthur, "Competing Technologies, Increasing Returns, and Lock-In by Historical Events," *Economic Journal*, 99.394 (1989), pp. 116–31; Liebowitz and Margolis, "Path Dependence, Lock-In, and History."

29 It is hard to invoke Hayek here after Slobodian's recent book *Hayek's Bastards* and Naomi Beck's rigorous exploration of Hayek's theories and their mistakes: Naomi Beck, *Hayek and the Evolution of Capitalism* (University of Chicago Press, 2018); Quinn Slobodian, *Hayek's Bastards: Race, Gold, IQ, and the Capitalism of the Far Right* (Zone Books, 2025). It is not surprising to see the "bastards" and their ideas in the 2020s – they were abundantly present, if completely outside the mainstream, in 1990s cyberculture.

30 A key text for the period was Huberman, "The Ecology of Computation." See especially the article Mark S. Miller and K. Eric Drexler, "Markets and Computation: Agoric Open Systems," in *Ecology of Computation*, ed. Huberman, pp. 133–75; and also Don Lavoie, Howard Baetjer, and William Tulloh, "High-Tech Hayekians," in *Markets, Information and Communication* (Routledge, 2003), pp. 107–41. Eric Raymond was also fond of Hayek in Eric S. Raymond, *The Cathedral & the Bazaar: Musings on Linux and Open Source by an Accidental Revolutionary* (O'Reilly, 1999). The convergence of these ideas about economics, complexity, and AI is represented by William J. Clancey, Stephen W. Smoliar, and Mark Stefik, *Contemplating Minds: A Forum for Artificial Intelligence* (MIT Press, 1994).

31 Timothy C. May, "Crypto-anarchy and Virtual Communities," MIT CSAIL Mailing List 1994. The documentary "Before the Web" documents the Hayekian subculture (https://reason.com/video/2020/10/07/before-the-web-the-1980s-dream-of-a-free-and-borderless-virtual-world). See also Salin, "The Wealth of Kitchens," an online essay with a nice bibliography of heroes: http://philsalin.com/kitchens/index.html.

32 Vikash Yadav, *Liberalism's Last Man: Hayek in the Age of Political Capitalism* (University of Chicago Press, 2023).

33 Friedrich A. Hayek, "The Use of Knowledge in Society," *American Economic Review*, 35.4 (1945); F. A. Hayek, *Law, Legislation, and Liberty: A New Statement of the Liberal Principles of Justice and Political Economy*, Vols. I–III (Routledge, 1982 [1973–9]).

34 Philip Mirowski and Dieter Plehwe, *The Road from Mont Pèlerin: The Making of the Neoliberal Thought Collective, with a New Preface* (Harvard University Press, 2015). See also the documentary *Chicago Boys*, dir. Carola Fuentes and Rafael Valdeavellano; as well as Medina, *Cybernetic Revolutionaries.*

35 On the Santa Fe Institute, Helmreich announced the death of artificial life we could have had in the space between a hardcover and a paperback edition of his book – see Helmreich, *Silicon Second Nature.* On Foresight and transhumanism, see Farman, *On Not Dying.*

36 Hayek, "The Use of Knowledge in Society," p. 519.

37 Beck, *Hayek and the Evolution of Capitalism*, p. 5.

38 "Abstract" society was Popper's name for the same problem: Popper, *The Open Society and Its Enemies*, 2 vols. Hayek dismissed Popper's ideas, but both of them took the world to be ordered by a radical split between small-group societies and modern, complex, urban – or, perhaps, global – social arrangements and economies.

39 Beck, *Hayek and the Evolution of Capitalism*, p. 4.

40 Hayek, *Law, Legislation, and Liberty*, Vol. I, p. 11.

41 Hayek, *Law, Legislation, and Liberty*, Vol. I, p. 17.

42 In volume I, pp. 1–2 of Hayek, *Law, Legislation, and Liberty*, Vol. I, p. 1.

43 John Perry Barlow, "A Declaration of the Independence of Cyberspace," *The Humanist*, 56.3 (1996), p. 18; Lawrence Lessig, "Reading the Constitution in Cyberspace," *Emory Law Journal*, 45.3 (1996), pp. 869–910.

44 Similar ideas animated folks looking at the internet and distributed computing in the 1980s and 1990s: Kevin Kelly's *Out of Control*; Rodney Brooks' work on robotics; Claus Emmeche's *Garden in the Machine.*

45 Hayek, *Law, Legislation, and Liberty*, Vol. I, p. 5.

46 Beck, *Hayek and the Evolution of Capitalism*, p. 8.

47 Nathan Schneider, *Governable Spaces: Democratic Design for Online Life* (University of California Press, 2024); Schneider, "Decentralization."

48 Ostrom, *Governing the Commons*; Charlotte Hess and Elinor Ostrom, *Understanding Knowledge as a Commons: From Theory to Practice* (MIT Press, 2007). Most Ostrom fans conveniently forget that she too was a neoliberal, of a certain sort – obsessed too with institutional redesign – married to Vincent Ostrom who was a president of the Public Choice Society.

49 Garrett Hardin, "The Tragedy of the Commons: The Population Problem Has No Technical Solution; It Requires a Fundamental Extension in Morality," *Science*, 162.3859 (1968), pp. 1243–8.

50 In legal terms, there was a robust literature in the 1990s on private orderings of various sorts – modes of governing that opted out of state sovereignty but reproduced parts of what it sought to provide: Robert C. Ellickson, *Order without Law: How Neighbors Settle Disputes* (Harvard University Press, 1991); Lisa Bernstein, "Opting out of the Legal System: Extralegal Contractual Relations in the Diamond Industry," in *Contract – Freedom and Restraint* (Routledge, 2000); Lisa Bernstein, "Private Commercial Law in the Cotton Industry: Creating Cooperation through Rules, Norms, and Institutions," *Michigan Law Review*, 99.7 (2001), pp. 1724–90.

51 James Boyle, *Shamans, Software, and Spleens: Law and the Construction of the Information Society* (Harvard University Press, 1996); Milton L. Mueller, *Ruling the Root: Internet Governance and the Taming of Cyberspace* (MIT Press, 2002); Laura DeNardis, *Protocol Politics: The Globalization of Internet Governance* (MIT Press, 2009); Kaushik Sunder Rajan, *Pharmocracy: Value, Politics, & Knowledge in Global Biomedicine* (Duke University Press, 2017); Silvia Lindtner, "Hacking with Chinese Characteristics: The Promises of the Maker Movement against China's Manufacturing Culture," *Science, Technology, & Human Values*, 40.5 (2015), pp. 854–79.

52 Jeff Juris, among others, was able to distinguish the finer points of how "networked sovereignty" might differ from spontaneous order: Jeffrey S. Juris, "The New Digital Media and Activist Networking within Anti-Corporate Globalization Movements," *Annals of the American Academy of Political and Social Science*, 597.1 (2005),

pp. 189–208; Jeffrey S. Juris, "Networked Social Movements: Global Movements for Global Justice," *Network Society: A Cross-Cultural Perspective* (2004), pp. 341–62. Michael Hardt and Antonio Negri briefly embraced a partial version of open source ideals and practices as part of arguing for new global social movements: Hardt and Negri, *Empire*.

53 E. Gabriella Coleman, *Hacker, Hoaxer, Whistleblower, Spy: The Many Faces of Anonymous* (Verso, 2015); Peter Ludlow, *Crypto Anarchy, Cyberstates, and Pirate Utopias* (MIT Press, 2001); Sarah Myers West, "Survival of the Cryptic: Tracing Technological Imaginaries across Ideologies, Infrastructures, and Community Practices," *New Media & Society*, 24.8 (2022), pp. 1891–1911.

54 Early observers of Amazon's power: Lilly C. Irani and M. Six Silberman, "Turkopticon: Interrupting Worker Invisibility in Amazon Mechanical Turk," in *Proceedings of the SIGCHI Conference on Human Factors in Computing Systems* (New York, NY, USA), CHI '13, 2013, pp. 611–20; Alessandro Delfanti, *The Warehouse: Workers and Robots at Amazon* (Pluto Press, 2021). Google's power was evident by this time as well: Zittrain, *The Future of the Internet*; Vaidhyanathan, *The Googlization of Everything*.

55 "Jonathan Postel," *IEEE Annals of the History of Computing*, 21.3 (1999), pp. 69–70.

56 Laura DeNardis, *Opening Standards: The Global Politics of Interoperability* (MIT Press, 2011); Russell, *Open Standards and the Digital Age*; Greene, "Landlords of the Internet."

57 Kelty, *Two Bits*.

58 Delfanti, *The Warehouse*; Brad Stone, *The Everything Store: Jeff Bezos and the Age of Amazon* (Little, Brown and Company, 2013).

59 Stone, *The Everything Store*, p. 210.

60 Stone, *The Everything Store*, p. 210.

61 Stone, *The Everything Store*, p. 222.

62 Schneider, *Governable Spaces*; Scholz and Schneider, *Ours to Hack and to Own*.

63 Luis Felipe R. Murillo and Christopher M. Kelty, "Hackers and Hacking," in *Digitisation: Theories and Concepts for Empirical*

Cultural Research, ed. Gertraud Koch (Routledge, 2017), pp. 95–116.

64 Indeed, see Peter Thiel, quoted in Fourcade and Healy, *The Ordinal Society*, p. 141.

65 Geert Lovink, *Sad by Design: On Platform Nihilism* (Pluto Press, 2019).

66 Bill Maurer, *How Would You like to Pay? How Technology Is Changing the Future of Money* (Duke University Press, 2015); Lana Swartz, *New Money* (Yale University Press, 2020).

67 Consider also recent writing about the "legitimation crisis" of Facebook and its oversight board: Gilad Abiri and Sebastian Guidi, "From a Network to a Dilemma: The Legitimacy of Social Media," *Stanford Technology Law Review*, 26 (2022), p. 92; Evelyn Douek, "The Meta Oversight Board and the Empty Promise of Legitimacy," *Harvard Journal of Law & Technology (Harvard JOLT)*, 37.2 (2023), pp. 373–446. See also the proposal for rethinking the study of platforms: Koray Caliskan, Donald MacKenzie, and Michel Callon, "Stacked Economization: A Research Program for the Study of Platforms," *Journal of Cultural Economy*, 18.2 (2024), pp. 304–31. Conversely, as I have been writing this book, the US government has taken ownership stakes in Nvidia and, through Oracle, in TikTok, among others.

68 Schneider, *Governable Spaces*; Tarnoff, *Internet for the People*; Taylor, *The People's Platform*.

If Participation . . .

1 On animation, see Teri Silvio, *Puppets, Gods, and Brands: Theorizing the Age of Animation from Taiwan* (University of Hawaií Press, 2019). She argues that a key feature of animation is the ratio of creators to characters – some animation involves people playing multiple roles, some involves multiple people animating a single character. The latter is at work in the animation of deliberation.

2 Arendt, *The Human Condition*; Mill, *On Liberty and Other Essays*; Jacques Rancière, *Disagreement: Politics and Philosophy* (University

of Minnesota Press, 1999); Chantal Mouffe, "Deliberative Democracy or Agonistic Pluralism?" *Social Research*, 66.3 (1999), pp. 745–58; Jürgen Habermas, *The Structural Transformation of the Public Sphere*, ed. Thomas Burger and Lawrence Kert (MIT Press, 1989).

3 Mill, *On Liberty and Other Essays*; Habermas, *The Structural Transformation of the Public Sphere*; Max Horkheimer and Theodor W. Adorno, *Dialectic of Enlightenment* (Continuum, 1994).

4 Nancy Fraser, "Rethinking the Public Sphere: A Contribution to the Critique of Actually Existing Democracy," in *Between Borders* (Routledge, 1994); Michael Warner, *Publics and Counterpublics* (Princeton University Press, 2021); Alexander Kluge and Oskar Negt, *Public Sphere and Experience: Analysis of the Bourgeois and Proletarian Public Sphere* (Verso Books, 2016).

5 Pierre Bourdieu, "Public Opinion Does Not Exist," in *Communication and Class Struggle*, ed. Armand Mattelart and S. Siegelaub (International General / IMMRC, 1979), p. 224; Thomas Osborne and Nikolas Rose, "Do the Social Sciences Create Phenomena? The Example of Public Opinion Research," *British Journal of Sociology*, 50.3 (1999), pp. 367–96; Mark G. Schmeller, *Invisible Sovereign: Imagining Public Opinion from the Revolution to Reconstruction* (Johns Hopkins University Press, 2016); John Durham Peters, "Concept of Public Opinion," *Public Opinion and the Communication of Consent* (1995), p. 1.

6 Benedict Anderson, *Imagined Communities: Reflections on the Origin and Spread of Nationalism* (Verso Books, 2006).

7 Guy Debord, *Society of the Spectacle*, rev. ed. (Black & Red, 1977).

8 Neal Stephenson, *In the Beginning ... Was the Command Line* (William Morrow & Co., Inc., 1999).

9 Craig Jarvis, "Cypherpunk Ideology: Objectives, Profiles, and Influences (1992–1998)," *Internet Histories*, 6.3 (2021), pp. 315–42; Enrico Beltramini, "Against Technocratic Authoritarianism: A Short Intellectual History of the Cypherpunk Movement," *Internet Histories*, 5.2 (2020), pp. 101–18.

10 Bruce Sterling, *The Hacker Crackdown: Law and Disorder on the Electronic Frontier* (Bantam, 1993).

11 Marcos, *Shadows of Tender Fury*; Claire Lisa Evans, *Broad Band: The Untold Story of the Women Who Made the Internet* (Portfolio/Penguin, 2018); Barlow, "A Declaration of the Independence of Cyberspace"; Peter Weibel and Timothy Druckrey, *Net Condition: Art and Global Media* (Steirischer Herbst, 2001). Arch private property theorist Milton Friedman lent his support to the Creative Commons license project in 2001.

12 The distinction between the internet and the World Wide Web is an easy thing to make too much of: the key effect of WWW was to bring the internet to widespread popular consciousness in 1993–4, but it neither created nor changed the general flexibility and extensibility of the internet at the time. If anything, it is part of a longer story of digital libraries: Christine L. Borgman, "Libraries, Digital Libraries, and Data: Forty Years, Four Challenges," *Portal: Libraries and the Academy*, 25.3S (2025), pp. 39–58.

13 Arendt, *The Human Condition*, p. 181. As for labor, Arendt's third category, she argues at length that the private sphere, the domain of labor, was elevated to the "social," creating an economic and physical arrangement of affairs where people are deprived of the power of speech and appearance. Something of this was visible here in that to experiment with the internet was to do so "on one's own time" – it was not for anyone yet a job, but the experimentation required the support of a job, or a family supporting the transcendance of that realm of necessity, in order to take place at all.

14 Boyle, *Shamans, Software, and Spleens.*

15 E. Gabriella Coleman, "Code Is Speech: Legal Tinkering, Expertise, and Protest among Free and Open Source Software Developers," *Cultural Anthropology*, 24.3 (2009), pp. 420–54.

16 Lessig, *Code.*

17 Kelty, "The Fog of Freedom."

18 Facemash was an immediate privacy disaster, first called out as such by Latina and Black women at Harvard: www.thecrimson.

com/article/2003/11/19/facemash-creator-survives-ad-board-the.

19 Finn Brunton, *Spam: A Shadow History of the Internet* (MIT Press, 2013); Jeong, *The Internet of Garbage.*

20 The distinction between pseudonymity and anonymity structured many early aspects of the internet we could have had – from the anonymous emailer anon.penet.fi and the early debates around the communication decency act: Gia B. Lee, "Addressing Anonymous Messages in Cyberspace," *Journal of Computer-Mediated Communication*, 2.1 (1996), p. JCMC211; Ludlow, *Crypto Anarchy, Cyberstates, and Pirate Utopias.* The stage was set early: anonymous, "The Constitutional Right to Anonymity: Free Speech, Disclosure and the Devil Notes and Comments," *Yale Law Journal*, 70.7 (1960), pp. 1084–1128.

21 Cf. "privately ordered law" and Scott's version of anarchism: Ellickson, *Order without Law*; Bernstein, "Opting out of the Legal System"; James C. Scott, *The Art of Not Being Governed: An Anarchist History of Upland Southeast Asia* (Yale University Press, 2009).

22 Julian Dibbell, "A Rape in Cyberspace or How an Evil Clown, a Haitian Trickster Spirit, Two Wizards, and a Cast of Dozens Turned a Database into a Society," *Annual Survey of American Law*, 1994.3 (1994), pp. 471–90.

23 Michael H. Goldhaber, "The Attention Economy and the Net," *First Monday*, 2.4 (April 1997).

24 From 1998 to 2005, digg, metafilter, advogato, techdirt, plastic.com, "Cool site of the day" (and Reddit, of course, who won) were each experiments in filtering and finding before social media algorithms appeared.

25 Mill, *On Liberty and Other Essays.*

26 Kelty, *Two Bits.*

27 Glyn Moody, *Rebel Code: Linux and the Open Source Revolution* (Basic Books, 2009).

28 Kelty, *Two Bits.*

29 Tim Berners-Lee, James Hendler, and Ora Lassila, "The Semantic Web," *Scientific American*, 284.5 (2001), pp. 34–43. The term Web

2.0 was sometimes used interchangeably with this term, but this is a mistake.

30 For those in computer science and information science, the relevant fields are data mining, user profiling, and other such approaches to observing traffic on the internet, and especially using web server logs. "Collaborative filtering" was one early name for using log data as a source of algorithmic sorting and recommendation.

31 Ada Lerner and others, "Internet Jones and the Raiders of the Lost Trackers: An Archaeological Study of Web Tracking from 1996 to 2016," in *Proceedings of the 25th Usenix Security Symposium (Usenix Security 16)*, Austin, Texas, August 2016.

32 The following sections rely heavily on Harry Halpin, "The Hidden History of the Like Button: From Decentralized Data to Semantic Enclosure," *Social Media + Society*, 9.3 (2023), pp. 1–13. Halpin goes into more detail about the specific histories, technology, and consequences of the semantic web and the Like button.

33 Carolin Gerlitz and Anne Helmond, "The Like Economy: Social Buttons and the Data-Intensive Web," *New Media & Society*, 15.8 (2013), pp. 1348–65. They noted "the Like economy is facilitating a web of positive sentiment in which users are constantly prompted to like, enjoy, recommend and buy as opposed to discuss or critique."

34 Stanley Milgram, *The Individual in a Social World: Essays and Experiments* (Addison-Wesley Pub. Co., 1977); Duncan J. Watts and Steven H. Strogatz, "Collective Dynamics of 'Small-World' Networks," *Nature*, 393.6684 (1998), pp. 440–2.

35 Leigh Ellen Gray, "Thumb War: The Facebook Like Button and Free Speech in the Era of Social Networking," *Charleston Law Review*, 7 (2012), p. 455.

36 See for instance, Taylor Lorenz's epic history of internet celebrity: Taylor Lorenz, *Extremely Online: The Untold Story of Fame, Influence, and Power on the Internet* (Simon & Schuster, 2023).

37 Is there a technical history of clickbait? If not, I guess this footnote will have to do. Maybe see Daniel Jácobo-Morales and Mauro Marino-Jiménez, "Clickbait: Research, Challenges and

Opportunities – a Systematic Literature Review," *Online Journal of Communication and Media Technologies*, 14.4 (2024), p. e202458.

38 Halpin, "The Hidden History of the Like Button," pp. 6–7.

39 Gerlitz and Helmond, "The Like Economy."

40 See US Patent #6285999, non-ironically available at https://patents.google.com/patent/US6285999.

41 Steven Levy, *In the Plex: How Google Thinks, Works, and Shapes Our Lives* (Simon & Schuster, 2011). Zuboff also makes use of this detail in Zuboff, *The Age of Surveillance Capitalism*.

42 Zittrain, for instance, was clear-eyed about some of the possible futures in 2008: Zittrain, *The Future of the Internet*.

43 Vannevar Bush and others, "As We May Think," *The Atlantic Monthly*, 176.1 (1945), pp. 101–8.

44 Taina Bucher, *If. Then: Algorithmic Power and Politics* (Oxford University Press, 2018); Kelley Cotter, "Playing the Visibility Game: How Digital Influencers and Algorithms Negotiate Influence on Instagram," *New Media & Society*, 21.4 (2018), pp. 895–913; Malte Ziewitz, "Rethinking Gaming: The Ethical Work of Optimization in Web Search Engines," *Social Studies of Science*, 49.5 (2019), pp. 707–31.

45 Nick Seaver, "Care and Scale: Decorrelative Ethics in Algorithmic Recommendation," *Cultural Anthropology*, 36.3 (2021), p. nil.

46 This is, quite literally, Zuboff's argument in *Surveillance Capitalism*: that there is no place called home anymore.

47 Cf. Fung and Cohen's late-to-the-party account of the virtuous citizen necessary to have a digital public sphere: Joshua Cohen and Archon Fung, "Democratic Responsibility in the Digital Public Sphere," *Constellations: An International Journal of Critical & Democratic Theory*, 30.1 (2023), pp. 92–7. A similar sentiment is visible in a review article by Henry Farrell, one of the best and earliest scholars of the internet in political science, which argued that eventually no one would need to study the internet because we would learn to account for any minor causal effects that it might have had on politics. I'm unsure whether this was an account of the internet we could have had, or one that dramatically under-

estimated the emergence of the one we do have: Henry Farrell, "The Consequences of the Internet for Politics," *Annual Review of Political Science*, 15.1 (2012), pp. 35–52.

48 Fraser, "Rethinking the Public Sphere"; Warner, *Publics and Counterpublics*; Nilüfer Göle, "The Gendered Nature of the Public Sphere," *Public Culture*, 10.1 (1997), pp. 61–81; Iris Marion Young, *Inclusion and Democracy* (Oxford University Press, 2002).

Else Hackers

1 Michael Schudson, *The Rise of the Right to Know: Politics and the Culture of Transparency, 1945–1975* (The Belknap Press of Harvard University Press, 2015).

2 Ithiel de Sola Pool, *Technologies of Freedom* (Harvard University Press, 1983), p. 1.

3 On hackers, see Sterling, *The Hacker Crackdown*; Coleman, *Hacker, Hoaxer, Whistleblower, Spy*; Luis Felipe R. Murillo, *Common Circuits: Hacking Alternative Technological Futures* (Stanford University Press, 2025); Christopher M. Kelty, "Every Era Gets the Internet It Deserves (or, the Phases of Hacking)," in *Exotic No More, 2nd Edition*, ed. Jeremy MacClancy (University of Chicago Press, 2019).

4 Beto O'Rourke notwithstanding. Hackers can become politicians, but generally not the reverse. Beto was a member of one of the most famous hacker collectives of the 1990s, the Cult of the Dead Cow. See Joseph Menn, *Cult of the Dead Cow: How the Original Hacking Supergroup Might Just Save the World* (PublicAffairs, 2019).

5 Lyotard, *The Postmodern Condition*, p. 103n231.

6 Murillo, *Common Circuits*; Kelty, "Every Era Gets the Internet It Deserves (or, the Phases of Hacking)."

7 Jodi Dean, *Publicity's Secret: How Technoculture Capitalizes on Democracy* (Cornell University Press, 2018).

8 E. Gabriella Coleman, "From Internet Farming to Weapons of the Geek," *Current Anthropology*, 58 (2017), pp. S91–S102.

9 See E. Gabriella Coleman and Christopher Kelty, eds., *Limn (Number 8): Hacks, Leaks and Breaches* (February 2017).

10 Matt Goerzen and E. Gabriella Coleman, *Wearing Many Hats: The Rise of the Professional Security Hacker* (Data and Society, 2022).

11 Coleman, *Hacker, Hoaxer, Whistleblower, Spy.*

12 E. Gabriella Coleman, "From Busting Cults to Breeding Cults: Anonymous H/Acktivism Vs. The (A)Nonymous Far Right and QAnon," *HAU: Journal of Ethnographic Theory*, 13.2 (2023), pp. 248–63 (p. 249).

13 Thi Nguyen explores the complexities of transparency and privacy in ways that aim at making sense of a similar but more general set of problems: C. Thi Nguyen, "Transparency Is Surveillance," *Philosophy and Phenomenological Research*, 105.2 (2021), pp. 331–61.

14 Patchen Markell, "Anonymous Glory," *European Journal of Political Theory*, 16.1 (2016), pp. 77–99; Ashley Gorham, "Anonymous's Glory," *International Journal of Communication*, 14.0 (2020), p. 19.

15 Coleman, "From Busting Cults to Breeding Cults."

16 Coleman, *Hacker, Hoaxer, Whistleblower, Spy*; Gorham, "Anonymous's Glory."

17 Joan Donovan, Emily Dreyfuss, and Brian Friedberg, *Meme Wars: The Untold Story of the Online Battles Upending Democracy in America* (Bloomsbury Publishing USA, 2022).

18 Coleman, "From Busting Cults to Breeding Cults," p. 256.

19 Susan Claudia Lepselter, *The Resonance of Unseen Things: Poetics, Power, Captivity, and UFOs in the American Uncanny* (University of Michigan Press, 2016), p. 156.

III Future Exception Handling

1 https://mukurtu.org.

2 Arturo Escobar, "Welcome to Cyberia: Notes on the Anthropology of Cyberculture," *Current Anthropology*, 35.3 (1994), pp. 211–31; Arturo Escobar, *Designs for the Pluriverse: Radical Interdependence, Autonomy, and the Making of Worlds* (Duke University

Press, 2018); *A World of Many Worlds*, ed. Marisol de la Cadena and Mario Blaser (Duke University Press, 2018); Tim Ingold, "One World Anthropology," *HAU: Journal of Ethnographic Theory*, 8.1–2 (2018), pp. 158–71; Martin Savransky, *Around the Day in Eighty Worlds: Politics of the Pluriverse* (Duke University Press, 2021).

3 See part four of Doctorow, *Enshittification*.

The Brains We Could Have Had?

1 Cf. Arendt, *The Human Condition*, pp. 171–2.

2 A concise version is in Evans, *Broad Band*.

3 Their names were: Kathleen McNulty Mauchly Antonelli; Jean Jennings Bartik; Frances (Betty) Snyder Holberton; Marlyn Wescoff Meltzer; Frances Bilas Spence; Ruth Lichterman Teitelbaum. They worked with John W. Mauchly and J. Presper Eckert, who for most of the twentieth century were known as the inventors of the ENIAC. See also Jennifer S. Light, "When Computers Were Women," *Technology and Culture*, 40.3 (1999), pp. 455–83.

4 As I write, a letter from a law firm in the matter of the $1.5 billion Anthropic settlement lies on my desk like the last will and testament of an internet we could have had, defrauded of its accumulated wealth, in its dementia and senility, by criminal scammers.

Acknowledgments

I am grateful to Dan Wielsch and the Wissenschaftskolleg of Berlin for inviting me to the meeting that occasioned the essay that became this book. I am grateful also to Geert Lovink for publishing that essay at the Institute for Network Cultures, one of the stalwarts of the internet we could have had. Kim Fortun read that essay and encouraged me, as she always does, and I trust her, so I kept writing. Sébastien Broca republished that essay in French and continued a long-standing conversation between us. Along the way a variety of people have offered similar much appreciated encouragement: Peter Stacey, Sarah Kareem, Eddo Stern, Daniel Snelson, Pamela Hieronymi, Luis Felipe Murillo Rosado, Anna Weichselbraun, Gabriella Coleman, Nick Seaver, Finn Brunton, Christina Dunbar-Hester, Marcos García, Nathan Schneider, Luke Stark, Stephen Collier, Andrew Lakoff, James Leach, Bill Rankin, Lisa Messeri, Joanna Radin, John Durham Peters, and the Yale Critical Computing Studies group. Three anonymous readers for Polity Press offered incisive encouragement, and John Thompson, Lindsey Wimpenny, Leigh Mueller, and others at Polity have done an amazing job of helping the manuscript become a book.

This book would not have been possible without the incredibly generous support of the Livescu Foundation, which funds the Initiative for Neuro, Narrative, and AI at UCLA; and in particular the generosity and support of Simona Livescu.

I owe the title of this book to Hannah Landecker, along with everything else precious to me.

Index